Environment, Health and History

Environment, Health and History

Edited by

Virginia Berridge
Professor of History, London School of Hygiene and Tropical Medicine, University of London, UK

Martin Gorsky
Senior Lecturer in the History of Public Health, London School of Hygiene and Tropical Medicine, University of London, UK

First published 2012 by
PALGRAVE MACMILLAN

Palgrave Macmillan in the UK is an imprint of Macmillan Publishers Limited,
registered in England, company number 785998, of Houndmills, Basingstoke,
Hampshire RG21 6XS.

Palgrave Macmillan in the US is a division of St Martin's Press LLC,
175 Fifth Avenue, New York, NY 10010.

Palgrave Macmillan is the global academic imprint of the above companies
and has companies and representatives throughout the world.

Palgrave® and Macmillan® are registered trademarks in the United States,
the United Kingdom, Europe and other countries.

ISBN 978-1-349-31322-8   ISBN 978-0-230-34755-7 (eBook)
DOI 10.1057/9780230347557

This book is printed on paper suitable for recycling and made from fully
managed and sustained forest sources. Logging, pulping and manufacturing
processes are expected to conform to the environmental regulations of the
country of origin.

A catalogue record for this book is available from the British Library.

Library of Congress Cataloging-in-Publication Data
Environment, health, and history / edited by Virginia Berridge,
Martin Gorsky.
p. cm.
Includes index.

1. Environmental health.   I. Berridge, Virginia.   II. Gorsky, Martin, 1956–
RA565.E473 2011
362.196'98—dc23   2011020967

10   9   8   7   6   5   4   3   2   1
21 20 19 18 17 16 15 14 13 12

Transferred to Digital Printing in 2012

Contents

List of Illustrations

List of Figures

List of Tables

Foreword

Prof. Sir Andy Haines
London School of Hygiene and Tropical Medicine

This book, which resulted from papers given at a landmark conference held at the London School of Hygiene and Tropical Medicine, provides a wide-ranging and authoritative overview of the evolution of our understanding of the interlinkages between environment and human health. While the relationship between the environment in which people live and their health has been known since the dawn of medicine it has often been consigned to the sidelines, particularly in relation to the contemporary emphasis on bio-medicine which focuses on individual relationships between health professionals and patients. Although this approach has resulted in substantial advances in the treatment and prevention of diseases, environmental factors currently play a major role in determining health, particularly in low-income countries. According to World Health Organisation estimates approximately one quarter of the global disease burden, and more than one third of the burden among children, is due to modifiable environmental factors.

Much of the earlier knowledge about environmental exposure and ill health arose from very high occupational exposures of workers in different industries to a range of chemical agents, and the subsequent development and application of epidemiological techniques to assess their impacts on health. With increasing regulation in the developed countries much of the industrial production which has the potential to result in such exposures has been moved to middle, and increasingly low-income countries with low health and safety standards. Likewise, partly because of the relative success of clean air legislation in developed countries, poor nations currently bear the brunt of indoor and outdoor air pollution, which are responsible for around 1.6 million and 800,000 annual deaths respectively. Increasingly therefore concerns about inequities in health and the influence of social determinants in maintaining such inequities are converging with the environmental health agenda.

There are also 'unfinished agendas' remaining from past ideological conflicts, such as the large numbers of nuclear weapons remaining from the Cold War. The failure to abolish nuclear weapons is fuelling nuclear proliferation as more nations perceive the political advantages to be gained from possessing weapons of mass destruction. The detonation of even a small proportion of existing nuclear weapons could create global environmental havoc, threatening crop production around the world as a result of the transport of large amounts of dust into the atmosphere which in turn would block sunlight from the Earth's surface and raises the potential spectre of mass starvation.

Although from our contemporary vantage point it may seem obvious that the existence and continued survival of humanity is dependent on the integrity of the planet's 'life-support' systems, including ecosystems that underpin food production and biodiversity and the availability of fresh water in the face of growing demands, this perspective has only recently begun to make its mark on academic discourse and public policy. Just as public health researchers and research funders have paid too little attention to the environmental determinants of health, environmental researchers, activists and policy makers have, all too often, been seemingly unaware of or indifferent to the interwoven relationships which bind together human health and environmental integrity. Increasingly it has become imperative to address these epistemological discontinuities to create genuinely transdisciplinary collaborative working which will illuminate our understanding of these complex relations and strengthen the evidence base for policies to provide the conditions for healthy living for a world population of 9 billion or more by the mid-century. Failure to do so, against a background of multiple challenges for humanity, including climate change, ocean acidification, depletion of fresh water resources, deforestation (especially of tropical forests), dramatic loss of species and burgeoning population growth (particularly in sub-Saharan Africa), will result in much of humanity being excluded from the health dividends arising from the dramatic social progress of recent decades. At the same time these challenges also pose wider questions for 'developed' societies whose trajectories of economic growth have been based on the unsustainable exploitation of affordable fossil fuels and other resources. This path of development will be increasingly denied to many 'developing' countries at a time when they are also suffering disproportionately from the impacts of global environmental change to which they have contributed little.

This book provides a range of invaluable perspectives on how the knowledge of the environmental influences on human health has developed over time. It illuminates the strengths and weaknesses of previous attempts to advance knowledge and the successes and failures of policies arising from environment and health research. It will prove a seminal guide to all those with an interest in this growing field of research.

Acknowledgements

This book began life as the 2007 conference of the European Association for the History of Medicine and Health held at the London School of Hygiene and Tropical Medicine. We are grateful to the Scientific Board of the Association for their support. We are also grateful for financial support from the Wellcome Trust, the London School of Hygiene and Tropical Medicine, and the Society for the Social History of Medicine. Colleagues at the School helped with organisation: Rachel Herring, Ingrid James, Alex Mold and Suzanne Taylor. In the production of this book, our thanks are due to Ingrid James for editorial support, to the anonymous referees who reviewed our proposal, and to Ruth Ireland and Michael Strang for support from Palgrave Macmillan. We are grateful to Diana LeCore for compiling the index.

We are grateful to Dr K. S. Hocking for permission to reproduce the photograph in Chapter 7. Images in Chapter 8 are published courtesy of the University of Strasbourg and of the National Library of Medicine, Bethesda, US. The photo of the El Paso smelter in Chapter 9 is courtesy of El Paso Public Library. This chapter is a shortened version of a paper in *Medical History* in 2010.

Notes on Contributors

Jane M. Adams completed her PhD in 2004 since when she has worked as a research fellow at the Centre for the History of Medicine at the University of Warwick. She is completing a manuscript provisionally titled 'Healing Waters: Spas and the Water Cure in Modern England' based on research funded by the Wellcome Trust. Her current research interests are alternative medical practice and the organisation of health services in nineteenth and twentieth-century Britain.

Virginia Berridge is Professor of History and Head of the Centre for History in Public Health at the London School of Hygiene and Tropical Medicine, University of London. She was president of the European Association for the History of Medicine and Health in 2007 and one of the organisers of the conference on the environment and health. Her published work is on substance use, HIV/AIDS, evidence and policy and post-war public health. Her most recent book is *Voluntary Action and Illegal Drugs* 2010 Basingstoke: Palgrave Macmillan (joint with Alex Mold) and she is currently working on an oral history of responses to swine flu.

Christian Bonah is Professor of the History of Medical and Health Sciences at the University of Strasbourg and currently holds a research professorship at the Institut Universitaire de France. He has worked on comparative history of medical education, the history of medicines and vaccines, as well as the history of human experimentation. Recent work includes research on risk perception and management in drug scandals and courtroom trials as well as studies on medical film.

Simon Carter is a lecturer in the Department of Sociology at the Open University. He has a particular interest in science and technology studies, especially as applied to issues of health and medicine. He has recently completed an historical study examining the cultural turn towards the sun and sunlight in early twentieth-century Europe and this research provides an analysis of the roles that sunlight played in the mediation of such notions as health, pleasure, the body, gender and class. He has also conducted research into critical approaches to the public understanding of science as applied to health issues.

Sabine Clarke is a Wellcome Trust Research Fellow and Lecturer in Modern History at the University of York. She is currently writing a book about the relationship between scientific research into industrial and medical uses of cane sugar and economic development plans for the British West Indies after 1940.

Martin Gorsky is senior lecturer in the contemporary history of public health and health systems in the Centre for History in Public Health at the London School of Hygiene and Tropical Medicine, University of London. His most recent book was *Mutualism and Health Care: British Hospital Contributory Scheme in the Twentieth Century* (Manchester University Press, 2006) (with John Mohan). His current research includes projects on the public health poster in Poland, the history of management in the NHS and the derivation of morbidity trends from sickness insurance records.

Andy Haines was Dean (subsequently Director) of the London School of Hygiene & Tropical Medicine between 2001 and 2010. In that role he was responsible for the management of over 1000 staff and 3700 postgraduate students. He was previously an inner-city GP in London and professor of primary health care at University College London. He is currently Professor of Public Health and Primary Care at the London School of Hygiene & Tropical Medicine.

Christopher Hamlin is Professor of History and of History and Philosophy of Science at the University of Notre Dame and Honorary Professor of Public Health and policy at the London School of Hygiene & Tropical Medicine. His most recent book is *Cholera: The Biography* (Oxford, 2009). Current projects include a history of fever, and studies of local nuisances administration and of disinfection policies.

Vanessa Harding is Professor of London History at Birkbeck College, University of London. She is interested in the interactions between people and the built environment, and has co-directed several related research projects on family, household, housing and health in early modern London. Her book *The Dead and the Living in Paris and London, 1500–1670* Cambridge: Cambridge University Press, appeared in 2002.

Ingar Palmlund is an independent scholar. After holding senior positions in the Swedish civil service, her academic affiliations have included Clark University, Tufts University, and the Fletcher School of Law and Diplomacy in the US, Linköping University and the Karolinska Institute in Sweden and the Wellcome Trust Centre for the History of Medicine at University College London in the United Kingdom. Her research concerns politics over the risks to human health and environment.

Lisa Rumiel is a Social Sciences and Humanities Research Council post-doctoral fellow at McMaster University in Hamilton, Ontario. Her research focuses on the intersection of health, the environment, nuclear technology, American militarism and social activism in historical context. She is currently working on transforming her dissertation – 'Random Murder by Technology: The Role of Physicians and Scientists in the Anti-Nuclear Movement, 1969–1992' – into a book and has begun research on a project about the role of University of Washington ecologists and biologists in the Pacific Proving Ground.

Dieter Schott is Professor of Modern History at the Technische Universitaet Darmstadt, Germany. His special fields of interest are the history of urban infrastructures and urban environmental history, particularly city-river relations. His publications include *Networking the City. Municipal Energy Policies, Public Transport and the Production; of the Modern City. Darmstadt Mannheim Mainz 1880–1918* (Darmstadt: Wissenschaftliche Buchgesellschaft 1999) and *Resources of the City. Contributions to an Environmental History of Modern Europe*, edited with Bill Luckin and Geneviève Massard-Guilbaud (Aldershot: Ashgate, 2005).

Christopher Sellers is an Associate Professor of History at Stony Brook University. He is the author of *Hazards of the Job* (University of North Carolina Press, 1997), as well as a forthcoming 'eco-cultural' history of post-World War II suburbanisation in the United States. He is currently studying the history of industrial hazards in the US compared to Mexico.

Christian Warren is Associate Professor of History at Brooklyn College. He is the author of *Brush with Death: A Social History of Lead Poisoning* (Johns Hopkins University Press, 2000), and co-editor of *Silent Victories: The History and Practice of Public Health in Twentieth Century America* (Oxford University Press, 2006). His next two books will explore why, how, and at what costs people have spent more and more time indoors.

Paul Wilkinson is Reader in Environmental Epidemiology, London School of Hygiene & Tropical Medicine. He trained in medicine and public health in Oxford and London, UK, and began epidemiological research at the National Heart & Lung Institute before moving to the London School in 1994. His principal research interests are climate and health; the health consequences of environmental change; and methods for assessing environmental hazards to health. He is a co-director of the London School's World Health Organisation Collaborating Centre on Global Change and Health.

List of Abbreviations

AEC	Atomic Energy Commission
AHRC	Arts and Humanities Research Council
BEIR	Biological Effects of Ionising Radiation
BHC	Benzene Hydrochloride
BHRA	British Health Resorts Association
BMA	British Medical Association
BSF	British Spa Federation
CEDR	Comprehensive Epidemiologic Data Resource
CFC	Chlorofluorocarbon
CH_4	Methane
CDW	Colonial Development and Welfare
CIC	Colonial Insecticides Committee
CO_2	Carbon Dioxide
CTB	Comprehensive Test Ban
DDT	Dichlorodiphenyltrichloroethane
DOE	Department of Energy
DSIR	Department of Scientific and Industrial Research
EMB	Empire Marketing Board
EPA	Environmental Protection Agency
FAO	Food and Agriculture Organisation
GHG	Greenhouse gas
HHS	Health and Human Services
HVAC	Heating, ventilation and air-conditioning
IEER	Institute for Energy and Environmental Research
IPCC	Intergovernmental Panel on Climate Change
ICRP	International Commission on Radiological Protection
ICSU	International Council for Science (formely International Council of Scientific Unions)

ILO	International Labour Organisation
INC	International Negotiation Committee
IPPNW	International Physicians for the Prevention of Nuclear War
ISMH	International Society of Medical Hydrology
ITC	International Tuberculosis Campaign
LSHTM	London School of Hygiene and Tropical Medicine
MAC	Maximum allowable concentration
MH	Ministry of Health
NEJM	*New England Journal of Medicine*
NHI	National Health Insurance
NHS	National Health Service
NRC	Nuclear Regulatory Commission
ppm	Parts per million
PSR	Physicians for Social Responsibility
SCOPE	Scientific Committee on Problems in the Environment
SPEERA	Secretarial Panel for the Evaluation of Epidemiological Research Activities
UFA	Universal Film Aktiengesellschaft
UN	United Nations
UNESCO	United Nations Educational, Scientific and Cultural Organisation
UNEP	United Nations Environment Program
UNFCCC	United Nations Framework Convention on Climate Change
USIS	United States Information Service
USPHS	United States Public Health Service
WHO	World Health Organization
WMO	World Meteorological Organization

1
Introduction: Environment, Health and History

Virginia Berridge and Martin Gorsky

The environment and health are not always considered in the same breath. In 2009, when the *Lancet* published a special issue on the impact of climate change on health, commentators noted that the health impact of global warming had come on to the agenda only since the 1990s.[1] Why are issues which appear to be closely connected, and have been connected in the past, so often discussed apart? This book historicises the changing nature of the connection, or lack of it, over time. It examines how and why health and the environment have been considered separately or together and why the relationship has changed. Such change has also taken place against a backdrop of evolving ideas of what is meant by 'the environment' and by 'health'. The traditional environmental rubric of 'airs, waters and places' has come to encompass the built environment and urban growth. Deforestation, the extension of irrigation and the conversion of grasslands came on to the agenda as the environment itself changed. What is meant by health, and the term itself, has also changed across the centuries. Most recently the reformulations of 'public health' have intersected with the 'new environmentalism', as we will see below. This introductory chapter is in three sections. First, we survey very broadly the relationship between health and the environment from the Greeks to the near present, with brief allusions to where the book chapters fit into that time frame; then we examine how historians have written about health and the environment; and finally, we summarise the arguments which our authors make in this book.

We can characterise the relationship in three broad time phases. From the time of the Greeks, environment and health concerns were entwined and this relationship was exemplified in the nineteenth-century public health movement. But then environment and health concerns drew apart in the early twentieth century; environmentalism emerged as

1

a separate movement which had little of a health dimension. In more recent times the two strands have begun to be considered together: the revival of environmentalism since the 1960s and 1970s has impacted on and helped determine new forms of public health.

1.1   From ancient Greece to the 1890s

The interconnection between health and the environment was strongly marked in ancient medicine: the tradition that the atmosphere and the environment might affect patterns of disease dates back thousands of years. Hippocrates in *Airs, Waters, Places*, in the fifth century BC, commented that physicians should have:

> due regard to the seasons of the year, and the diseases which they produce, and to the states of the wind peculiar to each country and the qualities of its waters ... the localities of towns, and of the surrounding country, whether they are low or high, hot or cold, wet or dry ... and of the diet and regiment of the inhabitants.[2]

The Hippocratic idea of the four humours (blood, yellow bile, black bile and phlegm); four qualities (hot, cold, dry and wet) and four seasons inextricably brought environment and health together. This, however, was a fatalistic doctrine; it accepted the connection but argued that it could not be modified.

The connection between health and the environment continued to be accepted much later – into the fifteenth and sixteenth centuries in the West. Disease and geography drew attention to the interconnection between malaria and marshy areas, a connection epitomised by the 'mal-air', in the name of the disease.[3] In fact, historians have drawn attention to the 'Columbian exchange' between the Old World and the New, where new environments were crucial to the spread of different diseases.[4] This was a dual exchange. Columbus and his explorers brought diseases such as measles to the New World; they, in turn, imported diseases like syphilis into Europe. A different form of transfer took place from the seventeenth century. This was a movement of scientific ideas. Historians such as Grove have argued that it was colonial scientists who recognised that the consequences of deforestation were soil erosion and shortage of water, and who were in essence developing a conservationist mentality.[5]

It was in the seventeenth century too that enquiry into the relationship between disease and place began, stimulating a new 'medical arithmetic' which recorded mortality alongside variables such as climate, air quality and elevation.[6] Theory initially moved cautiously from observed correlations to speculation about causes: were the agents miasmatic emanations or contagious particles, and what were the material or meteorological triggers for epidemics? Urban panic ultimately trumped lack of certainty, and preventive strategies were implemented in the burgeoning eighteenth-century cities, including drainage, cleansing of streets and stenches, and removal of burial sites.[7] **Harding** in our volume considers health and housing in this period. However, it was in the nineteenth-century urban public health movement that disease and environment became most closely entwined. Now, though, another variable was introduced, when in the 1820s the French statistician Villermé first detected the association between poverty, place and death rates.[8] The classic investigations of nineteenth-century living conditions therefore brought together not only accumulated beliefs about salubrity and hygienic practices, but also empirical study of the environment of poverty. Thus in Britain Edwin Chadwick's seminal survey began with the housing of the 'labouring classes', then moved to the streets outside before returning to 'internal economy' and 'domestic habits'.[9]

The main outline of the public health story is well known, through the rapid growth of cities from the late eighteenth century under the impact of industrialisation, the poor living and working conditions, and the epidemics of infectious disease which resulted.[10] The public health movement focussed on this poor environment as a cause of disease: miasmatic theories stressed the role of bad air and living and working conditions. The answer was seen in sanitation, in changing the environment and especially in providing clean water. In England the nineteenth-century public health acts (1848 and 1875, for example) focussed on the environment. These aimed at changing and regulating the environment; they emphasised clean water and the disposal of waste; housing; the regulation and improvement of working conditions and factories. Industrial legislation also combined health and the environment: the Alkali Acts of the 1860s were seminal for amenity and air pollution. Historians have argued about the rationale for such an emphasis within public health at that time. Clearly it was not simply altruism, but rather a fear of the urban mass and of contagion; the maintenance of economic advantage and of social cohesion drove the environmental focus. **Hamlin** has argued that public health took the path of the 'technical fix'. Sanitarianism, the doctrine of drains and clean water, precluded wider social reform, on the lines of the revolutionary political and health movements of 1848 in continental

Europe.[11] In this volume, Hamlin focuses attention on a non-industrial society at this period, that of famine era Ireland. What was the connection between environment and health there?

1.2 Environment and Health apart 1890s–1960s

From the end of the nineteenth century, ideas about public health and the environment moved apart. Science had much to do with it. The rise of bacteriology, the work of Koch and Pasteur from the 1880s and the rise of laboratory medicine, brought in their train what Starr has called 'a new concept of dirt'.[12] Worboys has shown that the new theories of disease were only gradually applied and there was considerable overlap with older ways of thinking, but they did focus attention on the role of the individual rather than of the wider environment.[13] Public health acts were no longer housing acts as well. Social and political events also underpinned this development. In the UK, the fear of degeneration of the race in the wake of the Boer War brought both positive and negative versions of eugenics, the positive focussing on the development of welfare systems, the negative on the removal of the unfit from society. The individual mother was often the centrepiece of welfare initiatives and there was stress on the role of personal hygiene. As Porter has commented, it was the domestic or inner environment that had to be changed rather than the wider environment; the element of personal culpability was stronger.[14] The separation was sometimes blurred: *Schott's* paper shows how meteorology continued to figure in German public health textbooks and the same was also true of standard British teaching texts.[15]

In the second half of the nineteenth century a trend emerged which dominated until the 1960s: for 'the environment' as a subject of interest developed rapidly and separately from health concerns. Most obviously this was through Romanticism – the work of Turner, Wordsworth, Rousseau and others and the fascination with 'wild places'. Intellectuals such as Morris and Ruskin looked back to medieval society to justify their rejection of economic liberalism. Their stance was tied to a 'back to the land' radicalism, exemplified by Henry George's *Progress and Poverty* and later by the Garden City movement. 'Problem solving' organisations concerned with the environment rapidly multiplied across Europe, typified in England by the Commons Preservation Society (1865), the Society for the Preservation of Ancient Buildings (1877) and the National Trust (1894). The movement, it has been argued, was divided between reformists and utopians. Should natural places be preserved for the people or from the people? There was no compromise with the industrial world and therefore little consideration of public health issues.[16] In fact the interests of environmentally inclined thinkers were often at odds with those of public health. Thirlmere reservoir in the Lake District was added to supplement Manchester's water supply in 1894 despite opposition from nature conservationists such as Morris, Ruskin and Octavia Hill.[17] Robert Roberts, in his recollections of an Edwardian childhood in the north of England, shows the existence of a popular environmental tradition as well.

> One sunny Wednesday afternoon, my mother took me to Peel Park. We sat on a high esplanade and looked far over the countless chimneys of northern Manchester to the horizon. On the skyline, green and aloof, the Pennines rose like the ramparts of Paradise. 'There!', she said, pointing, 'Mountains!' I stared, lost for words.[18]

Health figured little in these turn-of-the-century environmental movements. And later, in the interwar years, some aspects of environmentalism became discredited, in part because of a nascent connection with health. The environment became central to political ideologies, most notably those of Nazi Germany. The Nazis espoused both environmental and health promoting ideas, with a stress on 'back to the land' initiatives and on anti-smoking laws. Bramwell has called them 'the first radical environmentalists in charge of a state'.[19] That association with political extremism later helped to discredit such ideas until the revival of the 1960s. Schott shows the interconnections between public health ideas and environmentalism in the early twentieth-century period in Germany.

Elsewhere, the connection between environmentalism and health was on the wane at the state level or as a matter of public health concern. In the UK, environmentally focussed health treatments fell out of the mainstream (*Adams*); environmental movements such as the growing belief in the beneficial power of sunlight (*Carter*) were voluntary in nature. The revival of interest in the environment in the immediate post-war years had more connection with urbanisation and town planning than with health. In the UK, there was a concern to control environmental sprawl. Initiatives such as the Green Belt (1938 initially, just before the war); the Town and Country Planning Act (1947); and the introduction of national parks and access to the countryside (1949) were part of the post-war expansion of the state, but had little of a health component.

Meanwhile, public health both as an ideology and as a mode of professional practice went down a different route. In the UK, the legacy of the reforming doctrine of social medicine, which had considered the environment and had started to talk about the ecology of health, was twofold. One was a technical form of public health with a focus on chronic disease epidemiology and ultimately on the methodology of the randomised controlled trial (RCT). The other was health promotion, whole community ideas, which were to lead to a rapprochement between ideas about public health and those about the environment. In the short term however, the relocation of public health in England from local government to the National Health Service (NHS) at the local level resulted in the separation of formal public health professionals from the local environment. Environmental health and occupational health as areas with connections with the old nineteenth-century environmental vision, were left separate and took a different professional route.[20]

1.3   Environment and health reunited after the 1960s?

A new style of environmentalism emerged in the post-war years, which developed different forms of connection with health. In this collection chapters by *Clarke* and by **Bonah** show how the new post-war technocratic developments were used to understand and to portray the environment rather than being separate from it. The reconfiguration took different forms in different national contexts, as Clarke shows. In Britain, the 'great smog' of 1952 which led to the Clean Air Act of 1956 provides a model of post-war environmental interest with a health component. This looked both backwards and forwards – back to the environmentalism of nineteenth-century public health, but also forward to a new public health focussed on individual lifestyle and organised round single issue pressure groups.[21] The connection with smoking underlines the change which was taking place. The issue there was individual rather than a generalised industrial pollution. The Royal College of Physicians' first report on smoking published in 1962, was originally intended to deal with air pollution as well. But that subject was dropped; smoking was more easily characterised as a matter of individual responsibility, whereas air pollution raised much more difficult and general issues of industrial culpability.[22] In the US at the same period, as Sellers has shown, the reconfiguration of environmentalism was less about clean air and more about water and sanitation. The expansion of Long Island near New York and the lack of public health infrastructure there brought sanitation to the fore.[23] In different national cultures and institutional structures, the revival of environmentalism was differently configured. **Warren** discusses provocatively the impact of the post-war lifestyle changes and the move to a sedentary and indoor lifestyle.

Environmentalism in its recognisably modern form began to develop from the 1960s. Part of the rise of 'new social movements', it questioned, as did the anti-nuclear movement with which it had close connections, the role of science and technology. The oil crisis of the 1970s brought a realisation of the limits to economic expansion. Rachel Carson's *Silent Spring* (1962) was iconic, and the use of Agent Orange during the Vietnam War drew attention to the environmental impact of modern warfare. Ecology became a household word, and there was a realisation too about the health effect of pesticides on humans and on wildlife. The rise of the car was no longer seen as an automatic and liberating good. Buchanan's *Traffic in Towns* in the 1960s questioned the pro-car stance as did activism against lead in petrol in the 1970s. Environmentalism went mainstream with the establishment of environmental groups such as Greenpeace (1971) and Friends of the Earth (1971). Britain had its Department of the Environment (1970); the Green Party was set up in 1975. The campaign in 1971 to get Schweppes to drop non-returnable bottles marked a new style of direct action – media-conscious campaigning.

There was also a revival of the population concerns of the nineteenth century through the 'limits to growth' debate. Ehrlich's *The Population Bomb* (1968) was criticised for being more concerned about population growth in the Third World than in the US. There were underlying eugenic concerns with conservation issues too. Beinart and Coates have shown how conservationism could be authoritarian and not in tune with African countries' own concerns. White conservationists in South Africa were opposed by black people because of their intrusion into traditional settlement patterns.[24] This was environmentalism as a 'wilderness without people' rather than a lived experience.

Environmentalism became a global movement, and the interconnections with health also began to be reforged at that level. *Sellers'* paper illustrates this and shows the connection between local, national and international levels of action. This trend can be traced through the Club of Rome's *Limits to Growth* (1972); the UN World conference on the environment in the same year, and, in the 1980s, the Brundtland report, *Our Common Future* (1987). Environmental tragedies like Bhopal (1984) and Chernobyl (1986) drew attention to the global health effects of local environmental accidents. In this book, *Rumiel's* chapter shows how earlier medical anti-nuclear activism transmuted later into climate activism.

The new public health movement of health promotion brought health and the environment together. The Ottawa Charter of 1986 mentioned the two as part of the same continuum. The World Health Organisation's (WHO) Healthy Cities initiative, which began in 1986, exemplified the new interconnection. It operated at the local level and aimed to integrate health and the environment, consciously looking back to nineteenth-century public health.[25] Health and sustainability were discussed together at the Earth Summit in 1992 and through the activities of Agenda 21. Latterly health and climate change have been discussed. The chapter by **Palmlund** outlines these recent developments from her own committed perspective. Commentators have talked about the 'risk continuum' of recent public health: the environment generates risks but it is up to the individual to deal with them.

1.4   Historians, the environment and health

How have historians written about health and the environment? Historical work has tended to mirror the division between health and the environment which has been emphasised in the foregoing survey. Earlier work such as that by Crosby or McNeill's *Plagues and Peoples* took a macro approach, which did encompass the relationship between broad environmental changes and those in patterns of disease and mortality.[26] Otherwise, two schools of historical enquiry have established themselves, one on the environment and the other on health. These have only recently begun to consider each other and to interact. Environmental history over the last two decades has been a growth area of historical work. General surveys abound, as this subject has made its way onto student curricula.[27] But its concerns have been with the environment, often with an absence of people and certainly with an absence of health issues. The journal *Environmental History*, supported by the Forest History Society and the American Society for Environmental History, provides an example. A recent issue (January 2010) had papers on: Thoreau; the gender divide in conservation; deforestation; and the Atlantic alliance.[28] A recent major UK conference on the theme of 'environments' had just four papers that dealt with environment and health.[29] A special review of environmental history published to coincide with the conference covered ten books, none of which had a direct health focus.[30] Historians of environmentalism and conservation in Africa such as Beinart have drawn attention to the tensions between conservationist aims and those of indigenous populations; this divide has been replicated in the writing of environmental history, which has reinforced the absence of health concerns.[31] Mainstream environmental history has also mirrored the movement's own preoccupation with rural values, with little concern for the city.

Intersections with health history have developed in three areas: urban history; the history of occupational health; and the history of policy and social movements after 1945. Urban history initially figured little in histories of the environment. As Bill Luckin has commented, American environmental history concerned itself with the great West and the plains: 'the city loomed small in US environmental history'.[32] But later work such as Cronon's *Nature's Metropolis* (1991) and Melosi's *Sanitary City* (2000) began to bring both areas of interest together, as did Hamlin's history of UK public health in the Chadwick period.[33] The public health concern for pollution and the anti-smoke organisations which sprang up in Britain and the US by the end of the nineteenth century have been the subject of historical studies which draw together the connection between clean air and urban health.[34]

American environmental historians such as Sellers and Mittman have developed the tradition of occupational as well as public health history.[35] Their concern has been for the working environment and its impact on human health at local and global levels. Other historians have begun to look at the domestic environment, in a sense mirroring the way in which public health itself began to focus on that environment after the nineteenth-century heroic period.[36] A major environmental history initiative funded in the UK by the Arts and Humanities Research Council (AHRC) focussed on the theme of 'waste' and aimed to bring histories of health and of the environment together. The theme served to underline both the traditional forest and soil themes of mainstream environmental history and the role of waste as an issue at the local environmental level through recycling.[37] The most lively possibilities in recent work have been in the history of environmental campaigning and the rise of environmental issues in public health.[38] These dovetail with an interest in health history in 'new social movements' and post-war health campaigning.[39] But it is clear that there is much more that could be done to make the two historical strands of work aware of each other. Environmental history has been a multi-disciplinary enterprise, bringing together demographers, historians, epidemiologists and geographers. Epidemiologists have themselves been using historical data to answer contemporary questions.[40] One of the aims of this book is to integrate the work of health historians in this mix.

1.5   The contents of this book

The first chapters examine the changing relationship between environment and health in Europe at three different points in time: seventeenth-century London, Ireland during the great famine, and *fin de siècle* Germany. These were defining moments in the history of medical response to epidemic diseases. Plague-era London was where the intellectual foundations for the modern 'medicine of the environment' were laid, as Thomas Sydenham revived Hippocratic inquiry into the relationship between epidemics and place, and proto-demographers like William Petty and John Graunt first problematised the spatial patterning of mortality.[41] Ireland under the potato blight saw agricultural catastrophe trigger a mortality crisis of huge proportions, driven by malnutrition and infectious disease; yet the response was framed by political economy and medicine failed to articulate a new preventive agenda that linked environment, nutrition and health. By the end of the century an epidemiologic transition was underway in Western Europe, along with a paradigm shift in the science of disease causation, just as the technical achievement of urban sanitarianism was at its zenith. What would be the place of environment as 'bacteriological revolution' gained acceptance? In the contributions which follow the authors bring into view the material and medical contexts through which understanding of environment and health were refashioned during these pivotal phases.

Harding's study ranges beyond the established work of early modernists on the impact of plague.[42] Her method proceeds in three steps. First, she deploys the urban historian's techniques of reconstructing the physical presence of the city during its seventeenth-century expansion, examining its spread beyond the central parishes into the intra- and extra-mural suburbs. Alert to housing types, building layouts and landlord proclivity for subdivision she builds a picture of patterns of habitation and their accompanying social differentiation. Next she examines cause of death recorded in bills of mortality and parish records to reconstruct Londoners' sickness profiles, classifying diseases with an eye to subsequent spatial analysis. This leads her to posit three potential relationships between housing and health: poor water provision and sewerage as a cause of gastro-enteric disease, overcrowding as a factor in the spread of infection, and the impact of damp and poor building quality on the endemicity of 'consumption'. Drawing on qualitative sources she then records the perceptions of contemporaries about the health risks of dwelling in the 'Stinking parts of the city'. The opacity and fragmentary nature of the underlying sources render impractical the sort of regression analysis which would isolate the impact of housing from other causal factors, such as poverty and nutritional status. Nonetheless the broad conclusion is clear: a century of urban growth and the spread of poor-quality habitations coincided with high mortality from a range of diseases in which housing was implicated as a causal factor. This urban context was also one of the seedbeds from which the 'political arithmetic' of epidemiology would eventually grow.

While Harding's work treats the built environment as an unproblematic meeting place for urbanists and demographers, Hamlin's study emphasises awkward contradictions between the epistemologies of environmental and medical history. In thinking about health and disease the disciplinary traditions of the former incline towards the material and the *longue durée*, while those of the latter emphasise the political and economic contingencies that frame sickness and its responses. There could be no more poignant or contentious event through which to examine these interpretive tensions than the Irish famine of 1845–52. Hamlin begins by reflecting on the gulf between 'nature' and 'culture' in contemporary and subsequent readings of the famine. Historiographically this is a touchstone issue for nationalistic accounts, which depict the heavy death toll not as the outcome of agricultural disaster but of colonial callousness in land, trade and relief policies. Yet the famine was also a medical event in which deaths from infectious diseases probably exceeded those caused directly by malnutrition. Using his characteristic methodology of close reading of medical texts, Hamlin argues that in the pre-famine decades Irish doctors articulated a political medicine which addressed major questions of land policy and population health; at the same time their status made them arbiters of generous poor relief for 'fever' sufferers. Why then did the potential for a concerted medical response to the famine go unrealised? The reason lay with the reformed Irish poor law, which effectively subordinated doctors to bureaucrats in relief policy and choked off political medicine. This insight yields an important counterpoint to Hamlin's work on Britain in the same period, when doctors fell in behind Chadwick and resolved the problems of environment and health through the 'technical fix' of sewers and drains.[43] Similar possibilities beckoned in Ireland, through bog reclamation programmes to readjust the balance of population and cultivable land. But these never took off. Thus Hamlin's 'agroecological' account of the crisis gives full play to contingency while also remaining rooted in the biological.

Shifting perceptions of the interplay between environment and health at the cusp of the twentieth century are Schott's subject. Though nominally an investigation of the *Handbuch der Hygiene*, a multi-volume technical handbook begun in the 1890s, the study's agenda is to explore the complex influences exerted on public health by urban growth, intellectual currents in medicine and the broader cultural milieu. Today the meaning of the term 'hygiene' has narrowed from its classical sense, the preservation of good health, to evoke mere domestic cleanliness. In the context of the *Handbuch* though, it was a capacious trope. This was the moment when the new bacteriology challenged miasmatic theory and the environmental management it implied, yet these two ways of knowing (Koch versus Pettenkoffer, in the German context) sat uneasily alongside each other in the manual. Further volumes issued in the 1900s were to introduce 'social hygiene', signifying a move from population-level interventions to targeting risk groups. Thus far then, Schott illuminates from Germany's perspective the familiar progression of public health through its different modes, and he adds a further dimension to the work done by scholars such as Worboys in complicating the 'bacteriological revolution's' chronology.[44] More importantly, Schott's prosopographical reflection on the *Handbuch's* authors also permits him to recover the mental world of those who articulated these transitions. Loosely they manifested the Germanic 'progressivism' of an aspiring bourgeoisie, hitherto politically marginalised by Bismarckian conservatism. Their creed incorporated a technocratic and scientific response to the shock growth of second-wave industrial cities, melded with the incipient environmentalist and anti-urban critique of the 'Lebensreform' movement. These lifestyle- and nature-oriented campaigns will figure in later chapters too, in both British and American contexts. However Schott's argument here is quite specific: their German manifestation fed into a new aspiration to healthful city living in which hygiene was reinvented as a secular gospel of salvation. This turn might have laid the basis for an enduring environmental movement, as regions gradually addressed the challenge of urban pollution, and voluntary groups advocated the preservation of the natural landscape. Ultimately though, it foundered, because, Schott argues, here staking a position which is different to that of Bramwell's focus on Nazi environmentalism, depression and economic policy prioritised growth over environmental considerations.

If the turn of the twentieth century saw environmental matters diminish within public health, it also saw an intensification of popular interest in the health-giving properties of nature and the elements. In the manner of Schott's 'Lebensreform' activists, a host of groups emerged in Western nations to promulgate hiking, heliotherapy, fresh air, clothing reform, diet, exercise and so on.[45] This is perhaps para-doxical, given that the same period saw the building of health systems founded on curative biomedicine. The next section explores this phase, illuminating the contexts in which the healing powers of sun, airs and waters retained their place.

Carter's subject is sunshine, and that short, distinctive period when bodily exposure to solar rays was considered medically desirable. In discussing how this came about he tracks the development in Britain of cultural institutions and practices which idealised the tanned body and the outdoor life. As in Germany, the motive force was a reaction against the perceived degeneracy of urban industrial society, and the manifestations took different forms. There were the Boy Scouts, whose texts extolled a hardy masculinity distinguished by bronzed skin, and there were 'back to nature' advocates of camping, who sought to nourish both physique and psyche through the recapture of Arcadian innocence. Carter's concern though, is to locate these tendencies within broader public health discourses, in which heliotherapy gained a degree of medical respectability. They appealed particularly to the interests of social hygienists, who delineated the role of place and class in patterns of mortality, and incorporated eugenicist ideas about hereditary deter-minants of the ills of urban society. Solar therapy therefore became allied to the larger project of improving the housing and living condi-tions of city dwellers, which incorporated slum clearance and smoke abatement as well as the great outdoors. Its advocates appear at the same time visionary environmentalists, perceiving the need for clean energy in place of polluting coal, and creatures of a moment in moder-nity when liberal urban governments sought to control and make vis-ible the lives of 'degraded' and unhealthy citizens.

Where Carter emphasises the intellectual threads of anti-urbanism and eugenics which ran through the 'physical culture' movements, Adams stresses their embeddedness in leisure and tourism. Her subject is the water cure and its associated venue, the British spa town, for which the interwar years represented a last flourish before decline. Hydrotherapies had enjoyed a long history, with the spa reaching its apogee in the late-Victorian period, when, alongside the time-honoured ingestion of heal-ing waters, a panoply of other therapies were adopted. The bracing or soothing qualities of different spas were determined by considerations of climate and water quality, and clients enjoyed new modes of bathing using jets and showers, alongside dietary and exercise regimens created by medical hydrologists. Why then did this form of healing wane with the advance of state medicine? Adams traces the growth of the spa as commercial entity, with advertising designed to attract middle-class consumers away from seaside towns. Therapeutic relief of stress and nervous disorders was packaged alongside sports facilities, pleasant public spaces and sophisticated shopping opportunities. This though was a prospectus that sat uneasily with the concurrent economic strategy, of attracting working-class visitors by tapping into statutory funding under the new National Health Insurance scheme. The spas' bid for incorporation into state medicine turned on the claim that they could speedily return rheumatism sufferers to productive work, but this was ultimately rejected by government. Adams notes that although the given reason was cost containment, class prejudices were implicated too, as doubters feared the new clientele would spoil the refined atmosphere of the resort. Thus although spa treatments persisted in the early years of the NHS they were superseded by physiotherapy and movement-oriented hydrotherapy. These shifted the emphasis onto bodily exercise, in place of the holistic combination of climate, water and place.

In the next section the discussion turns away from the industrialised nations, to parts of the world where conquest of infectious disease through control of environment remained a focus of policy. New directions had been heralded by the disciplines of tropical medicine which established themselves during the high tide of European empire. Scientists like Manson, Laveran, Ross and Bruce had explicated the insect vectors of diseases such as filariasis, trypnasomiasis and malaria.[46] In their wake came entomologists, parasitologists and helminthologists for whom a new environmental health was implied, and one which was heavily determined by colonial imperatives. Thus the site of control moved from the city streets to the rural spaces on which the agricultural productivity of empire depended. Early interventions included both vector management strategies like bush clearance and marsh drainage, and parasite eradication efforts, such as Jamot's 'atoxylisation' of sleeping sickness patients in Tanzania. By the mid-century, with the end of empire imminent and with 'colonial welfare' eliding into 'development', a more ambitious programme emerged: eradication through large-scale chemical warfare against insects.

The subject of Clarke's chapter is the mass spraying of Dichlorodiphenyltrichloroethane (DDT) against malaria, and her purpose is to recover an episode in British environmental science which has hitherto been absent from the historiography. This has tended to emphasise the globalised nature of the campaign in the hands of the WHO, and to assume that blind faith in the agrichemical solution effectively set back critical research in malariology. Clarke's story refines and revises this view by uncovering the field's vigorous research culture in Britain in the 1940s and 1950s. Her case study allows her to unpick the network of interests which converged to drive this science: imperial economics, military capacities, academic and applied research institutes, and chemical manufacturing amongst others. The ensuing alliance of malariologists and the Colonial Office, lubricated by government funding for development, engendered a research programme that interrogated various aspects of eradication in the field. Much of this was practical: devising optimal techniques for aerial spraying or assessing the efficacy of insecticides other than DDT. However, some projects were evaluations of specific interventions, whether through systematic monitoring of reinfestation rates, health outcomes or economic analysis. What this reveals is understandings which were far removed from the hubristic faith in technological panacea which is sometimes attributed to the WHO campaign. Indeed, by the late-1950s, British researchers had articulated the fundamental paradox of the environmental strategy: the apparent impossibility of completely eradicating insect populations in even limited areas meant that sustained heavy spraying was needed. But this was both financially unviable and risked breeding resistance.

While a critical science had begun to undermine technocratic confidence by the mid-century, the same period also saw that faith affirmed in media representations. Bonah's concern is the depiction in film of environmental management as health intervention. His contention is that cinematic sources can reveal much of the mentalities which underpinned this strategic shift in tackling infectious diseases in colonial or low-income settings. The argument builds on several case study movies produced with industrial, military, governmental or international sponsorship, whose subject matter encompassed not only DDT but the earlier technique of anti-larvicidal oil spraying, as well as pest control. At one level this material is helpful in recording different applied techniques and revealing the various interests concerned with public information and education. However, Bonah goes further in exploring connoted meanings within the films, embedded in the language, imagery and staging of these essays in chemical triumphalism. The first point to emerge is the potency of the military metaphor, and this focuses attention both on the direct importance of war in furnishing technologies of eradication, but also in constructing environment as the habitat of dangerous vectors which must be targeted and destroyed. Allied to this was a vision of 'chemical modernity', premised on the liberating power of science and incorporating population health within the broader narrative of reconstruction. Thus behavioural approaches to malaria prevention such as bednet use and pharmacotherapies were sidelined and earlier physical interventions by drainage and reclamation gave way before the chemical fix. And the 'dreams' of eradication also obscured the part played by economic underdevelopment in sustaining the disease.[47]

In the final section the contributors bring contemporary historical perspectives to bear on issues which have emphatically reunited environment and health in the West. They also have weighty political currency: the hazardous materials we touch or inhale; our sedentary domestic lifestyles and their deleterious health effects; and, looming over all this, the prospects of nuclear winter or climate-change disaster. Linking these papers is a concern with that same nature/culture dichotomy which Hamlin's chapter identifies as the great challenge facing medico-environmental historians. Our instincts are to historicise, to set, say, anti-nuclear campaigners or global warming activists within their social and cultural frame, and to treat the 'science' as text, as contingency. Yet as scholars of environment, and as historians in the civic realm, can we really disregard the empirical and material as we contemplate these issues? Whether explicitly or implicitly, this group of authors pin their colours to 'nature' as their intellectual foundation, and develop arguments which unashamedly pack a moral punch.

Sellers takes as his starting point the necessity of incorporating contemporary understandings of the toxicity of dangerous substances in historical analysis. His subject is the health risks of lead, both to workers and to consumers, and he urges that the globalised economy demands of medical historians a fundamental rethink. The paradigm of 'occupational health' history proves to be too Western and too oriented to labour politics to be applicable now. On the one hand the production process has increasingly been externalised, removed to cheap, lightly regulated locations where it is invisible to consumers. On the other, there is complacency about the risk from lead now that atmospheric pollution is resolved, so that dangers associated with familiar commodities such as painted toys go unseen. Sellers argues instead for a cross-national approach to analysis, which connects local experiences of work hazards with global patterns of consumption, via national regimes of surveillance and measurement. He explores this through discussion of smelting works on the US/Mexico border serviced by Mexico's lead mines. In Texas this initially escaped the attention of interwar hygienists, thanks to its peripheral position and American distaste for international regulatory trends. After 1945 though, the combined disciplinary expertise of engineers, epidemiologists and toxicologists legitimised notions of 'safe' levels of lead. Meanwhile prosperity and suburbanisation encouraged an ideal of the 'clean' US factory, physically separated from the workforce. Just over the border though, risky production was stepping up to fuel world demand, as Mexican workers thronged to industrial locations where development imperatives meant occupational health was a low priority. We need to acknowledge our common histories and interdependencies, Sellers suggests, if we are to build genuinely responsible policy in this area.

While occupational health has been a constant in the history of public health, the relationship between housing and wellbeing has slowly disappeared from twentieth-century Western narratives. Warren's chapter seeks to overturn complacency about the risks which accompany indoor dwelling. His case is that however advantageous the spacious, hygienic shelter of the home had been in the eras when infectious disease menaced, it has become increasingly perilous over the last hundred years. The story begins in the early twentieth century, with American and British critics of the domestic lifestyles which cut off humanity from the more wholesome natural world. In part this is familiar terrain for readers of earlier chapters, with the postures of characters like John Harvey Kellogg not too far removed from Schott's 'Lebensreformers' or Carter's helio-advocates. Yet Warren heeds the possibility that they were far-sighted Cassandras glimpsing future public health threats. The case unfolds in three stages. First, the real physical risks of life inside are documented through a case study of rickets, whose resurgence he links not to the familiar issue of dietary deficiency, but to the in-dweller's inadequate exposure to sunlight. Second, he considers the onward march of the climate controlled environment, and while acknowledging some health benefits (air-conditioning as shelter from heat risk), he sees dangerous auguries for human adaptation in our growing intolerance of temperature extremes. Third, he explores the growing social isolation that accompanies the retreat into a private life dominated by electronic media. What chance now to accumulate the social capital so necessary for mental well-being and the diminution of health inequalities? Some readers may find in Warren's predictions of an enfeebled post-human future a contemporary iteration of the anti-urbanism articulated a century before by champions of the outdoors life. Alternatively, his essay may just be an early map of the new agenda for environmental history, soon to become an urgent priority.

The risks which Rumiel addresses are much more familiar: the threat to human well-being posed by military and civilian uses of nuclear power. She presents case studies of two physician organisations which emerged in the 1960s and 1980s in opposition to the threat of atomic war, and which are explicable as manifestations of Cold War politics. Why though did this campaign transmute to a more broadly based opposition to nuclear energy? Rumiel postulates that one answer was the signal influence of the Chernobyl disaster. The Soviet reactor meltdown dramatised safety issues and provoked anxieties for populations living in proximity to generators. It also galvanised a similar confluence of expertise to that noted by Sellers in the realm of lead exposures, with public health doctors now standing alongside other scientific disciplines in assessing the costs and benefits of nuclear energy. The question which this provokes is what place public health criteria should have in an environmental arena dominated by physicists and engineers, and shaped by economic and strategic concerns. The engagement of medical protagonists had derived initially from a sense of professional duty, but enthusiasm alone was not enough to ensure effectiveness. To be heard at the table, physicians needed to capitalise on events and to build bridges with other experts to convey their message.

At the time of writing, public concern about the safety of nuclear energy has emphatically revived following the 2011 Japanese tsunami and the Fukushima disaster. Now though this debate is couched within the larger context of the climate change challenge. Our collection closes with Palmlund's history of the international politics driving today's overriding environmental concern. Like Rumiel, her project is to track the place of public health in this policy discourse, and she too finds it disappointingly marginal in the risk accounting. She first provides a brief chronology of the arrival of climate science in the political arena, from the early transnational awakenings over ozone depletion, to the now global awareness of the effect of greenhouse gases. Parallel to this, international organisations arose to sustain the scientific consensus and broker negotiations about emissions reductions. Meteorology, agriculture and alternative energy technologies loomed large in these new networks, and where social impacts were discussed the emphasis was squarely on growth economics and prospects for development. Where then was health? Following preliminary warnings in the professional journals of a range of risks, such as heat wave mortality, the resurgence of vector-borne diseases and the stress on health systems of extreme weather events, the WHO belatedly entered the field. Palmlund though warns against any false optimism. The natural sciences still dominate the debate, she argues, and doctors have barely begun to model the effects and plan public health responses.

With environmental reform increasingly stalled by industry sceptics, North/South hostilities and the backsliding of short-termist politicians, Palmlund's committed reading suggests that now, more than ever, a new political medicine or public health is called for. The histories recounted in this collection counsel caution when contemplating the prospects for this. Whether in famine-era Ireland or the malarial regions of colonial Africa, science has to make its way among competing interests, and is indeed constructed by them. Yet the environmental imperative has traditionally been the core narrative of public health, whose articulation in Rosen's classic account began with the drainage systems of Mohenjo Daro and reached a dramatic climax with 'Enter Mr Chadwick'.[48] As Charles Webster has argued, public health advocates then gave a powerful example in facing down vested interests and elevating environmental health from a position of neglect to the centre of politics.[49] In recent times, that historical legacy had been adapted in the public health field – 'sanitarian becomes ecologist', in John Ashton's words.[50] The nineteenth century was important for the interaction of environment and health, but that relationship, so our book both argues and illustrates, has a wider and shifting set of historical relationships which also need to be taken account of in framing future strategies.

Notes

1. A. Haines, A. J. McMichael, K. R. Smith, I. Roberts, J. Woodcock, A. Markandya, B. G. Armstrong, D. Campbell-Lendrum, A. D. Dangour, M. Davies, N. Bruce, C. Tonne, M. Barrett and P. Wilkinson (2009). 'Public health benefits of strategies to reduce greenhouse-gas emissions: overview and implications for policy makers', *Lancet*, 19 Dec, 374 (9707), 2104–14. Epub 26 November 2009 review. A. Markandya, B. G. Armstrong, S. Hales, A. Chiabai, P. Criqui, S. Mima, C. Tonne and P. Wilkinson (2009). 'Public health benefits of strategies to reduce greenhouse-gas emissions: low-carbon electricity generation', *Lancet*, 12 Dec, 374 (9706), 2006–15; P. Wilkinson, K. R. Smith, M. Davies, H. Adair, B.G. Armstrong, M. Barrett, N. Bruce, A. Haines, I. Hamilton, T. Oreszczyn, I. Ridley, C. Tonne and Z. Chalabi (2009). 'Public health benefits of strategies to reduce greenhouse-gas emissions: household energy', *Lancet*, 5 Dec, 374 (9705), 1917–29. Epub. 26 November 2009.
2. Hippocrates (translated by Francis Adams), *On Airs, Water, and Places*, the *Internet Classics Archive* http://classics.mit.edu/Hippocrates/airwatpl.html, accessed 3 August 2010.
3. M. Dobson (1997). *Contours of Death and Disease in Early Modern England* (Cambridge: Cambridge University Press).
4. A. Crosby (2003). *The Columbian Exchange: Biological and Cultural Consequences of 1492* (Westport, Conn: Greenwood Press, 30th anniversary edition).
5. R. Grove (1996). *Green Imperialism: Colonial Expansion, Tropical Island Edens and the Origins of Environmentalism* (Cambridge: Cambridge University Press).

6. J. C. Riley (1987). *The Eighteenth-Century Campaign to Avoid Disease* (London: Macmillan); A. Rusnock (2002). *Vital Accounts: Quantifying Health and Population in Eighteenth Century*, (Cambridge, Cambridge University Press).
7. Riley, *The Eighteenth-Century Campaign to Avoid Disease*; M. Foucault (1974). 'The birth of social medicine', in J. Faubion (ed.) (1994). *Michel Foucault: Power. Essential Works of Foucault 1954–1984, Volume Three* (London: Penguin), pp. 135–56.
8. A. La Berge (1992). *Mission and Method: The Early Nineteenth-Century French Public Health Movement* (Cambridge: Cambridge University Press), pp. 54–75.
9. E. Chadwick (1842). *Report on the Sanitary Condition of the Labouring Population of Great Britain*, ed. M. W. Flinn (1965) (Edinburgh: Edinburgh University Press).
10. A. Hardy (1993). *The Epidemic Streets: Infectious Disease and the Rise of Preventive Medicine, 1856–1900* (Oxford: Clarendon Press).
11. C. Hamlin (1998). *Public Health and Social Justice in the Age of Chadwick. Britain, 1800–1854* (Cambridge: Cambridge University Press).
12. P. Starr (1982). *The Social Transformation of American Medicine* (New York: Basic Books), p. 189.
13. M. Worboys (2000). *Spreading Germs: Disease Theories and Medical Practice in Britain, 1865–1900* (Cambridge: Cambridge University Press).
14. D. Porter (1999). *Health, Civilisation and the State. A History of Public Health from Ancient to Modern Times* (London: Routledge), p. 143.
15. W. Wilson Jameson and F. T. Marchant (1920). *Hygiene, Specially Intended for those Studying for a Diploma in Public Health* (London: J and A Churchill).
16. See, for example, C. F. Mathis, 'Environmental campaigning in the nineteenth century', paper given at the 79th Anglo American conference of historians July 2010.
17. H. Ritvo (2009). *The Dawn of Green: Manchester, Thirlmere and Modern Environmentalism* (Chicago: University of Chicago Press).
18. R. Roberts (1976). *A Ragged Schooling: Growing up in the Classic Slum* (Manchester: Manchester University Press), p. 16.
19. A. Bramwell (1994). *The Fading of the Greens. The Decline of Environmental Politics in the West* (New Haven and London: Yale University Press); A. Bramwell (1989). *Ecology in the Twentieth Century: A History* (New Haven: Yale University Press); R. Proctor (1999). *The Nazi War on Cancer* (Princeton: Princeton University Press).
20. J. Lewis (1986). *What Price Community Medicine? The Philosophy, Practice and Politics of Public Health since 1919* (Brighton: Wheatsheaf Books).
21. R. Parker (1986). 'The struggle for clean air', in P. Hall, H. Land, R. Parker and A. Webb (eds). *Change, Choice and Conflict in Social Policy* (Aldershot: Gower), first published 1975 and reprinted 1978, pp. 371–409; V. Berridge and S. Taylor (2005). *The Big Smoke: Fifty Years after the 1952 London Smog. Proceedings of a Witness Seminar* (London: Centre for History in Public Health). http://www.ccbh.ac.uk/witness_bigsmoke_index.php, accessed 6 June 2011.
22. V. Berridge (2007). *Marketing Health. Smoking and the Discourse of Public Health in Britian, 1945–2000* (Oxford: Oxford University Press), p. 63.
23. C. Sellers (2006). 'Worrying about the water; sprawl, public health and environmentalism in post World War Two Long island', seminar given at Centre for History in Public Health, LSHTM 7 April.

24. W. Beinart and P. Coates (1995). *Environment and History: The Taming of Nature in the USA and South Africa* (London: Routledge); W. Beinart and L. Hughes (2009). *Environment and Empire* (Oxford: Oxford University Press); W. Beinart (2008). *The Rise of Conservation in South Africa* (Oxford: Oxford University Press).

25. V. Berridge, D. A. Christie and E. M. Tansey (eds) (2006). *Public Health and the 1980s and 1990s: Decline and Rise?* (London: Wellcome Centre at UCL). http://www.ucl.ac.uk/histmed/publications/wellcome_witnesses_c20th_med/vol_26, accessed 6 June 2011.

26. W. H. McNeill (1977). *Plagues and Peoples* (Oxford: Basil Blackwell).

27. I. Simmons (2008). *Global Environmental History* (Edinburgh: Edinburgh University Press); J. McNeill *Something New Under the Sun: An Environmental History of the Twentieth-Century World* (New York: W. W. Norton and Co); S. Mosley (2010). *The Environment in World History* (London: Routledge).

28. *Environmental History* (2010). 15(1) January.

29. Programme of the 79th Anglo-American conference of historians 'Environments' July 2010.

30. See *Reviews in history* <reviews-in-history> at http://www.history.ac.uk/reviews/subject/history-type/environmental-history, accessed 28 July 2011.

31. Beinart and Hughes, *Environment and Empire*; Beinart, *The Rise of Conservation*.

32. B. Luckin (2006). 'Environmental activism, environmental history' paper given to Centre for History in Public Health seminar LSHTM, 19 January. See also D. Schott, B. Luckin and G. Massard-Guilbaud (eds) (2005). *Resources of the City: Contributions to an Environmental History of Modern Europe* (Aldershot: Ashgate).

33. W. Cronon (1991). *Nature's Metropolis: Chicago and the Great West* (New York: W. W. Norton); M. Melosi (2000). *The Sanitary City: Urban Infrastructure in America from Colonial Times to the Present* (Baltimore: Johns Hopkins University Press).

34. S. Mosley (2008). *The Chimney of the World: A History of Smoke Pollution in Victorian and Edwardian Manchester* (London: Routledge); B. Luckin (1986). *Pollution and Control: A Social History of the Thames in the Nineteenth Century* (Bristol; Adam Hilger); A. Wohl (1983). *Endangered Lives: Public Health in Victorian Britain* (London: Methuen).

35. G. Mitman, M. Murphy and C. Sellers (eds) (2004). *Landscapes of Exposure: Knowledge and Illness in Modern Environments. Osiris* special issue, 19; P. Weindling (1986). *The Social History of Occupational Health* (London: Croom Helm/Society for Social History of Medicine).

36. M. Jackson (ed.) (2007). *Health and the Modern Home* (London: Routledge); C. Mills (2007). 'Coal, clean air and the regulation of the domestic hearth in post war Britain', in Jackson (2007); S. Mosley (2003). 'Fresh air and foul: the role of the open fireplace in ventilating the British home, 1837–1910', *Planning Perspectives*, 18, 1–21; Mosley, 'The Home Fires: Heat, Health and Atmospheric Pollution in Britain,1900–45', in Jackson (ed.). *Health and the Modern Home*, pp. 196–223.

37. T. Cooper (2008). 'Challenging the "refuse revolution": war, waste and the rediscovery of recycling', *Historical Research*, 81 (214), 710–31.

38. See Berridge and Taylor (2005). *The Big Smoke*; Berridge, Christie and Tansey (2006). *Public Health*.

39. See for example, J.-F. Mouhot, 'Environmental NGOs and environmental campaigning in Britain since 1945', paper given at the 79th Anglo-American conference.
40. K. G. Kuhn, D. Campbell-Lendrum, B. Armstrong and C. R. Davies (2003). 'Malaria in Britain: past, present and future', *PNAS*, 100(17), 9997–10001; C. Carson, S. Hajat, B. Armstrong and P. Wilkinson (2006). 'Declining vulnerability to temperature related mortality in London over the twentieth century', *American Journal of Epidemiology*, 164, 77–84; T. McMichael (2001). *Human Frontiers, Environments and Disease: Past Patterns, Uncertain futures* (Cambridge: Cambridge University Press).
41. Riley, *The Eighteenth-Century Campaign to Avoid Disease*, pp. 1–19, at 9.
42. P. Slack (1985). *The Impact of Plague in Tudor and Stuart England* (London: Routledge & Kegan Paul).
43. Hamlin, *Public Health and Social Justice*.
44. G. Rosen (1958). *A History of Public Health*, (edn 1993, Baltimore: Johns Hopkins University Press), chs vi, vii; Worboys *Spreading Germs*.
45. See inter alia: Porter, *Health, Civilisation and the State*, 303–9; J. Matthews (1990). 'They had such a lot of fun: The Women's League of Health and Beauty between the wars', *History Workshop Journal*, 30, 22–54; I. Zweiniger-Bargielowska (2007). 'Raising a nation of 'good animals': the new health society and health education campaigns in interwar Britain', *Social History of Medicine*, 20, 73–89.
46. D. M. Haynes (2001). *Imperial Medicine: Patrick Manson and the Conquest of Tropical Disease* (Philadelphia, University of Pennsylvania Press).
47. R. Packard (2007). *The Making of a Tropical Disease: A Short History of Malaria*, (Baltimore: Johns Hopkins University Press), pp. 175–6.
48. Rosen, *History of Public Health*, pp. 1, 175.
49. C. Webster (1992). 'Public health in decline', *healthmatters*, 11, 10–11.
50. J. Ashton (1991). 'Sanitarian becomes ecologist: the new environmental health', *British Medical Journal*, 26 January, 302, 189–90.

2
Housing and Health in Early Modern London

Vanessa Harding

The relationship between housing and health in early modern London is not easy to distinguish from the general relationship between wealth or poverty and health. Poor housing was only one of many disadvantages potentially compromising the health of the urban poor: it can be assumed that they had poorer diets, less adequate clothing, less access to clean water for cooking and washing, less access to medical treatment, and probably greater exposure to workplace dangers, than their richer contemporaries, even before we begin to look at their accommodation. And 'poor housing' itself has many facets, both physical and use-related: disadvantageous location, exposure to industrial and other pollution, quality of building materials, adequacy of repair, sanitation and water supply, heating and ventilation, furnishing, neighbourhood character, density and character of occupation. Nevertheless, there appears to be some correspondence between poorer areas of London and poor health outcomes; this paper seeks to explore some of the correlations and variables in that relationship.

It was certainly widely believed by contemporaries that early modern London was both absolutely and comparatively an unhealthy place to live. The mortality gap between city and country was made explicit by John Graunt in his *Natural and political observations upon the bills of mortality*, first published in 1662 and reprinted several times, itself an index of interest in the subject.[1] He noted that whereas one in 50 died in the country, one in 30 did so in London, approximating to modern estimates of 20 per 1000 in the country and 33 per 1000 in the city.[2] But although he made a number of suggestions as to why this should be so, he did not explore in any depth variations in mortality across the metropolis.[3]

Modern research shows that these were in fact complex and changing over time.[4] While there is a broad correlation between areas with higher levels of poverty, however indexed – parishes that were net beneficiaries of citywide poor rates, for example, or areas with high levels of exemption from taxes – and areas with high mortality rates, there was also considerable local variation and anomaly. Some research suggests that city-centre health improved, at least compared with the health of the extramural parishes, between the mid-sixteenth and the mid-seventeenth centuries, despite increasing population density. Paul Slack's study of plague incidence shows the focus of heavy mortality moving from within to outside the walls between 1563 and 1665, while Justin Champion's maps of mortality in the mid-seventeenth century show lighter endemic and epidemic mortality in most of the intramural (city-centre) parishes than in most of the extramural, despite the fact that the city centre was probably still more densely settled than the inner suburbs.[5]

But can we pick out the specific contribution of housing quality to the health of London's inhabitants? Looking for a connection between particular houses or areas of housing and particular health outcomes appears to require a specificity of information that in general is not available, as well as to assume that most individuals lived long enough in one property for its effect on their health to be significant. Comparatively few parish registers offer information on cause of death, and none does so for very long periods, so while very local studies may be possible, comparisons over time and space are difficult. So far we have had to proceed more by inference and hypothesis, using a range of qualitative as well as quantitative sources. The following discussion draws on research for the project 'Housing environments and health in early modern London, 1550–1750', funded by the Wellcome Trust, 2006–8, and the two related projects, 'People in place: families, households and housing in early modern London, 1550–1720' (funded by AHRC, 2003–6), and 'Life in the suburbs: health, domesticity and status in early modern London' (funded by ESRC, 2008–11).[6] 'Housing environments and health' aimed at relating mortality data (seasonality, age-group specific mortality and, where available, causes of death) to an integrated matrix of family reconstitutions and reconstructed property histories so that mortality could be located and mapped at the level of streets and houses. Most of the outputs of this project are yet to come, so this paper sketches the background to the enquiry rather than reporting detailed results.[7]

To attempt this, the paper offers a broad characterisation of housing types and areas, and a brief outline of early modern London's mortality profile, and then considers whether we can directly associate variations in health with features of housing quality or with the topography of poverty and poor housing. Specifically the discussion focuses on three themes: overcrowding, sanitation and building quality.

2.1 Housing quality and type

The pattern of housing development in early modern London is quite complex, but a broad division into three areas of housing type can help to provide an overview for the purposes of this study.[8] Early modern London grew outwards from a dense medieval centre, to become a sprawling metropolis engulfing several formerly distinct settlements. Following the divisions of the contemporary Bills of Mortality, we can contrast the city within the Roman-medieval walls ('the 97 parishes'), with the immediate extramural suburbs ('the 16 parishes'), and the outer suburbs including Westminster.[9] A distinctive group of housing types predominated in each of these broad areas, though all areas contained a variety of types. All three areas shared the same building traditions and materials, with timber, brick, and plaster as the main construction materials and clay tiles for roofing; what varied were the size and spatial disposition of properties, the quality of materials and workmanship in their construction, and the access, amenities and services available. The sources for understanding London housing in the early modern period include drawn and written surveys made for property owners and records of property management and use, such as leases and repair agreements or accounts.[10] Litigation over boundaries, amenities and nuisance can also be illuminating.[11]

In the 97 parishes within the walls, high land values, the constraints of space, and the demand for accommodation close to the centre encouraged development of better-built, taller houses. Houses of five or six stories lined the most desirable street-frontages, although in side streets two or three stories were more common. All houses of this kind would have had brick-built chimneys with hearths on several floors, and glazed or shuttered windows. Some were partly cellared, in brick or stone; some had separate kitchen outbuildings, others a kitchen within the main structure, and in general there were rooms for distinct functions, working, cooking, eating and entertaining, sleeping. Such houses were more likely to have a piped water supply and their own privy or privies, with a sealed cesspit. They might have a yard, or access to leads or a roof walk; a few of the most spacious had gardens.[12] Landlords had an interest in maintaining the quality of their city-centre premises in order to uphold the rents, and were less likely to divide them for multi-occupation, or to allow noxious work processes and practices to take place there. Population densities increased over the sixteenth and seventeenth centuries, at least up to the Great Fire of 1666, but only to about double their late-medieval level. There was a broadly concentric distribution of housing quality, declining towards the city walls where houses were smaller and cheaper.[13] Along the waterfront warehousing, brewing and other commercial activities seem to have diminished environmental quality, so that large old houses had often been divided into tenements or smaller dwelling units, and rent values were lower.[14]

The inner suburbs, here identified with the ring of 16 parishes surrounding the city wall, wholly or partly within the jurisdiction of the mayor and aldermen, were a rather different story from the more prosperous and well-built centre. Partly settled in the Middle Ages, main streets were lined with good houses, including many inns for the country traffic, but they were backed by a spread of newer and poorer housing. The capital's rapid early-modern population growth was concentrated in these areas between the mid-sixteenth and the mid-seventeenth centuries.[15] The new building pattern was constrained by the lines of major streets and the existing framework of freehold plots. Though space was cheaper and more available than in the centre, most inner-suburban housing was unable to command high rents and development was driven by economy rather than investment. Despite the city's and the crown's proclamations, standing houses were subdivided and gardens, backyards and open spaces were built over with cheap, low-rise housing. The most characteristic development was the alley or close, branching off main streets in a pattern like a fish's backbone. At least some of these alleys seem to have developed when a landowner or leaseholder occupying the street-front property either allowed his tenants to build houses on the back part of his plot or built them himself.[16] The Hearth Tax assessments of the 1660s and 1670s indicate that the majority of houses in the inner suburbs were small, with three or fewer hearths per house; only in more prosperous and well-inhabited streets on radial roads, especially to the west of the walled city such as Aldersgate Street, were there significant numbers of larger dwellings.[17]

While the better properties, perhaps belonging to an institutional landlord, were regularly surveyed and repaired, houses in the inner suburbs tended to be cheaply and often shoddily built – some were no more than converted outhouses and stables and were altered and subdivided in piecemeal fashion. Areas where regulation was limited or absent, such as Tower Hill in the Liberty of the Tower, were particularly bad.[18] Even the better houses were unlikely to have more than two or three rooms per family, and the occupants of the alleys shared their water supply and privies with other neighbours. The inner suburbs were also the site of manufacturing and processing industries, such as brewing, soap-boiling, and sugar-refining, which must have contributed to pollution as well as to local employment.[19]

Building in alleys and closes could not provide accommodation fast enough to house London's rapidly growing population, and a secondary form of suburban development took place, largely in the seventeenth century, in London's outer suburbs. These outer suburbs comprised both a cluster of West End parishes with some urban characteristics, and a more far-flung range of very large parishes from Clerkenwell to Stepney, and Bermondsey to Lambeth. Development in these areas was based not so much on existing main roads, though ribbon development did feature, but on property ownership or control: large landowners or developers created entirely new streets and street patterns, in order to penetrate and open up blocks of land, often hitherto green fields.[20]

Covent Garden is an early and important example. The site, originally a garden belonging to Westminster Abbey, had no frontage to the Strand, and could only be exploited for the landowner's benefit by positive planning. The Earl of Bedford obtained the necessary licence, with some difficulty and under numerous conditions, and laid out the square or piazza with associated streets in the 1630s. Although most of the houses were constructed on building leases rather than by Bedford, there was an overall plan and sufficient building control to ensure general uniformity of appearance, accommodation and facilities.[21] Perhaps more typical is what happened on the St Paul's estate in Shadwell, where groups of houses and whole streets were laid out as units from the 1620s, but with less of an overall plan. There were 703 houses by 1650 on what had been open ground 30 years before.[22] This pattern was repeated in other new areas, giving, no doubt, something more of uniformity to their appearance and character than was the case in older-established areas where ownership was fragmented and development took place in a more piecemeal fashion.

But the outer suburbs did not produce a single house type; already the different demands of East and West End economies were affecting the sorts of dwelling built there. The fashionable West End produced several different styles of building. Very large houses lined the Strand and High Holborn, with nine or ten rooms each, and smaller ones, with five or six rooms each, along Long Acre and Piccadilly. The older houses along the Strand were timber-framed, but the newer ones elsewhere were largely of brick.[23] The mean number of hearths per household in many western suburban parishes was five or more.[24] In the eastern outer suburbs, houses were generally smaller. Shadwell houses, like many in the inner suburbs, were commonly two-storey buildings with four rooms and garrets, built of timber and infilled with brick and daub. The mean number of hearths per house was three or fewer, but 40 per cent had gardens.[25] The outer suburbs also housed new large workplaces and commercial and industrial premises, some of which may have been inimical to health. An alum works in Wapping, for example, established in 1626, provoked numerous complaints and petitions for its removal.[26] The production or processing of naval stores and equipment dominated the eastern suburb from the Victualling Office by the Tower eastwards, while Spitalfields became a centre for the manufacture of silk and silk textiles.[27]

This broad characterisation of London's housing types maps onto the distribution of wealth and poverty in the seventeenth century. A survey of the rentable values of city properties in 1638 indicates a concentration of wealth in the most central parishes within the walls, where the median rent per annum was £10 and above, surrounded by poorer intramural parishes with a median rent of less than £10. In the parishes outside the walls, median rents were £10 or less, except towards the west, influenced by the growing attractions and prosperity of the new West End.[28] The Hearth Tax returns of the 1660s and 1670s document both the distribution of property size, as noted above, and also the prevalence of poor households (listed as exempt from the tax on account of poverty) in the outer suburbs, especially towards the east.[29] Returns for the valuations and taxes of the 1690s confirm this picture, with mean household rent values in the central and western parishes mostly exceeding £17, and those in the eastern and northern suburbs mostly less than £10. The range of variation was much greater in the western parishes, however, suggesting pockets of poverty close to some of the wealthiest properties.[30]

Significantly for London's overall wealth and health profile, a far higher proportion of its population was accommodated in the inner and outer suburbs by 1700 than had been the case in 1550. Between c. 1550 and 1700, the population of the 97 parishes within the walls increased from some 40–45,000 to 75,000; that of the inner suburbs from 26–30,000 to 170,000; and that of the outer and distant suburbs from 13–15,000 to over 300,000.[31] Even though parts of suburban London, especially towards the west, included areas of high value and high-quality housing, the majority of the increase was of much more modest dwelling types. Inevitably, therefore, this meant that more Londoners, and a higher proportion of the whole, lived in areas of generally poor housing quality. The consequential environmental benefits of the Fire of 1666 – rebuilding, street widening, the creation of dedicated marketing and quayside areas – were mostly confined to the city within the walls and a small part of the western suburb.[32] The northern, eastern and outer suburbs may all have felt the burden of extra population in the years following the Fire, increasing pressure on residential accommodation and services.

2.2   The topography of death and disease: the evidence of the London Bills of Mortality

Mapping patterns of mortality in early modern London depends to a great extent on the data captured in the annual Bills of Mortality, and on subsequent versions of that data.[33] The London Bills of Mortality distinguished plague and non-plague deaths from the outset, and listed causes of death other than plague from 1629. Despite the problems with translating the 'Diseases and casualties' identified in the bills into modern diagnoses, and the fact that many 'casualties' describe symptoms or coincidental phenomena rather than causes, it is possible to use the bills to generalise about London's mortality profile in the seventeenth century. Figure 2.1 presents some summary trends in the composition of deaths by major causes at three periods, based on Graunt's collection of cause-of-death data from 1629 to 1659 and the eighteenth-century printed 'Collection' of the yearly bills from 1657 to 1758.[34] There were three major components to this profile: infant mortality, epidemic mortality including plague, and endemic mortality including natural ('aged') deaths. Accidental and felonious deaths occurred in less significant numbers.

One of the most significant elements in London's mortality profile was the poor survival of babies and infants. Roger Finlay calculated mortality rates of between 167 and 265 per 1000 for children aged 0–4 years, in the later sixteenth to early seventeenth centuries, while Graunt observed in 1662 that about one-third of all deaths were from diseases of infancy.[35] Rates varied across the city, with the outer suburban parishes doing better than the crowded inner-city ones until the mid-seventeenth century, but with poorer inner-city and riverside parishes doing worst of all.[36] The situation may well have deteriorated rather sharply in the later seventeenth century, with infant mortality in the outer suburbs reaching levels similar to those in the inner city,

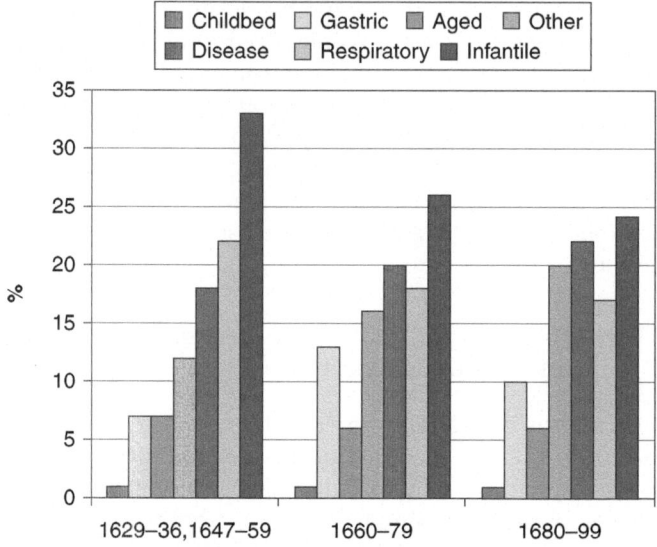

Figure 2.1 Causes of death (percentage distribution) in seventeenth-century London
Source: J. Graunt (1662). *Natural and political observations mentioned in a following index, and made upon the bills of mortality* (London); T. Birch (ed.) (1759). *A Collection of the Yearly Bills of Mortality, from 1657 to 1758 inclusive …* (London).

Key

Generic cause of death	'Causes' in the Bills of Mortality
Childbed	Childbed
Gastric	Flux, bloody flux, scouring, griping in the guts
Aged	Aged
Other	Chronic ailments (gout, palsy, etc.); accidental and felonious deaths; external or superficial symptoms (abscesses, ulcers, skin complaints); internal dysfunction (dropsy, rupture, stone, strangury); pox, syphilis
Disease	Ague, fever, typhus, purples and spotted fever, smallpox, measles, swinepox, jaundice, etc.
Respiratory	Consumption, cough, tissick, pleurisy, etc.
Infantile	Abortive, stillborn, overlaid and starved at nurse, convulsions, teeth, etc.

which were also rising. More recent family reconstitution and analysis suggests that infant mortality alone (deaths under one year old) was exceeding 300 per 1000 in around 1700.[37] It is difficult to disentangle some of the variables contributing to this. Inner-city parishes certainly sent numbers of their infants to the outer suburbs and the country to be wet-nursed, but they also received an influx of foundlings, whose chances of survival were lower than average. There may also have been a shift towards employing wet-nurses in the household by the later seventeenth century, thus increasing the city's population of nurslings, at risk of death within the city.

Early modern London was subject to a number of epidemic diseases occurring sporadically and sometimes very locally, and creating sharp fluctuations in mortality.

Plague, London's most famous killer disease, may have been present in London in two years out of three, but only a handful of deaths occurred in most years. The overall pattern is one of few if any deaths in most years, punctuated by extreme epidemic years in which tens of thousands of plague deaths were recorded. Plague epidemics occurred episodically, at intervals of between ten and twenty-odd years, and even in the mildest of these total mortality for the year doubled or trebled. In the worst plague years, arguably 1563 or 1625, deaths reached five or six times normal. As London grew, so did the total of plague deaths, so that even if 1665 was not the most severe plague in relation to the capital's total population, the numbers dying were much greater than in any previous epidemic. Epidemics usually died down very quickly, with few plague deaths the following year, except that the plagues of 1603 and 1636 were each followed by up to a decade of raised plague mortality.[38]

Other 'epidemicall' diseases ('*Purples, Spotted-Feaver, Small-Pox,* and *Measles*', according to Graunt) had different cycles and recurrences, at times coinciding with plague.[39] Unlike plague, however, these diseases killed significant numbers in most years, with regular peaks, suggesting they were both endemic and epidemic.[40] Smallpox was present in mid-seventeenth-century London, and seems to have established itself as both endemic and epidemic by the 1650s, with a three- to four-year cycle. It worsened appreciably in the 1680s, with a succession of years of high mortality, reaching nearly 13 per cent of all deaths in 1681, but death totals and proportions were less severe, around 5–6 per cent, by 1700.[41] Typhus (purples and spotted fever) was similarly variable but much less destructive, averaging 1 per cent of non-plague deaths.[42] Although Graunt listed 'ague and fever' among 'chronical' or endemic diseases, they are better considered here. Probably comprising both influenza and malaria, the category 'ague and fever' was a major killer, causing around 13.5 per cent of non-plague deaths over most of the seventeenth century. Severity varied over a five- or six-year cycle, from 9 or 10 per cent of deaths to 17–18 per cent in individual years.[43]

However, Graunt argued that it was not epidemic but '*Chronical* distempers' that were the key to whether a place was healthful or not. 'The *Chronical* Diseases shew the ordinary temper of the Place, so that upon the proportion of *Chronical* Diseases seems to hang the judgment of the fitness of the Country for *long Life*. For, I conceive, that in Countries subject to great *Epidemical* sweeps men may live very long, but where the proportion of the *Chronical* distempers is great, it is not likely to be so; because men being long sick and always sickly, cannot live to any great age'.[44] Among the '*Chronical* Diseases [...] whereunto the city is most subject', he cited 'for Example, *Consumptions, Dropsies, Jaundice, Gowt, Stone, Palsie, Scurvy, rising of the Lights, or Mother, Rickets, Aged, Agues, Feavers, Bloody-Flux,* and *Scowring*'.[45] Some of these were clearly infectious, some not, and they were of very differing significance in the mortality pattern, but as a group (apart from 'ague and fever'), they were comparatively stable from year to year, a constant strong undercurrent. Two elements of this group to be considered in more detail below are enteric/gastric complaints and respiratory ailments, comprising respectively the bills' categories of griping or plague in the guts, bloody flux, scouring, and vomiting, and consumption, cough, 'tissick', pleurisy and quinsy.

2.3   Housing and health

Slack's and Champion's work indicates, as already noted, that city-centre health improved vis-à-vis the health of the extramural parishes, between the mid-sixteenth and the mid-seventeenth centuries, and this would fit with a scenario of spreading poor-quality housing outside the walls, accommodating rapid population growth but entailing poor environmental quality. Landers too argues that a widening differential in housing and environmental quality between richer and poorer areas from the later seventeenth through the eighteenth century correlated with varying levels of mortality instability.[46] There would seem to be three main areas of likely correspondence between housing and health in early modern London: the density of human occupation, questions of water and sanitation, and issues of warmth and shelter. Of these, the first contributed to the spread of epidemic disease, but it must also have promoted ordinary infections and non-fatal complaints.

2.3.i   Overcrowding and the spread of disease

One key to the unhealthiness of the inner suburbs appears to have been the way people actually lived in the buildings. Elizabethan proclamations against suburban development argued that crowding of poor people contributed to the spread of disease, if not to its generation:

> where there are such great multitudes of people brought to inhabite in small roomes [...] and they heaped vp together, and in a sort smothered

with many families of children and seruantes in one house or small tenement, it must needes followe (if any plague or popular sicknes shoulde by Gods permission, enter amongst those multitudes) that the same would not onely spread it selfe and inuade the whole Citie and confines, as great mortalitie shoulde ensue to the same, [...] but woulde be also dispersed through all other partes of the Realme.[47]

This belief may well have been supported by the deaths of 'well near an hundred' people in the King's Bench prison in 1579, from 'a certain Contagion, called, The Sickness of the House', 'engend[ered] chiefly, or rather only, of the small or few Rooms, in respect of the many Persons abiding in them; and there, by want of Air, breathing in one anothers Faces as they lay, which could not but breed Infection'.[48] The pejorative term 'pestering' or 'pestered', of houses and alleys, used in city proclamations from the 1570s to indicate dividing into small tenements, became a standard language of topographical description, used by John Stow and others.[49] But the repetition of both royal and civic proclamations against dividing houses and new building in the suburbs is clear evidence that it continued.

Population density is more complicated than person per acre and has to be considered in the context of occupied floor space rather than the footprint of the buildings. City-centre parishes were very densely occupied, perhaps as high as 230 persons per acre in the very centre just before the Fire, grading down to 150 per acre or fewer just inside the walls. The inner suburbs had densities of 140–55 per acre.[50] (The outer suburban parishes were generally so large and so irregularly settled, with some large spaces remaining unbuilt, that it is difficult to give meaningful numbers). But city-centre houses had several floors and different rooms to accommodate their heavier population; the inner suburbs were populated by division and multi-occupancy and the rapid colonisation of any open space.

Landlords' attitudes were crucial to this. In the city centre (where a good deal of property was owned by city livery companies) there seems to have been a moral and economic motive to maintain properties at a high standard to attract high-paying tenants. In the new-planned suburbs such as Covent Garden and even Shadwell, an improving landlord could afford to invest in services appropriate to the area. But the intermediate inner suburbs seem to have been at the mercy of small profit-taking private landlords, who divided houses, accommodated inmates and built rows of small dwellings. Nor were institutional landlords exempt from subdivision.[51]

In the 1630s the Privy Council's anxiety about the extent of development resulted in an enquiry into divided houses across the city, the returns to which shine a searching light on recent developments.[52] In Portsoken Ward, viewers reported for instance that 'These nine tenements were but two till Ambrose Lenning wheelwright divided them and are now continued so by Richard Vauke stonecutter landlord'.[53] Landlords seem to have been unscrupulous in cramming in tenants: Katharine Floyd or Flood, herself living in the western suburb near St Sepulchre's church, was landlady to a property in Covent Garden precinct of Portsoken, 'a most poor close house and but three rooms in it', which she had divided into three units occupied in all by eight adults and eight children. She 'will in no waies be reformed in this abuse' and had 'other like houses which now stand voyde wch she intends to fill again when she hath patcht them up'.[54] A contemporaneous survey of the poor in Portsoken Ward identified 925 persons, 530 adults (a number of them aged or unable to work, but many willing but unable to get work) and 395 dependant children. Many of them lived in alleys or rents, thus confirming the association of alley properties with settled poverty.[55]

The 1630s were barely the midpoint of a century of rapid population growth and building development in the inner suburbs. Once an alley was established the houses might be further divided to provide more units of accommodation. Three Kings Alley in Aldgate, which seems to have been created between 1557 and 1584, was described in 1632 as 13 houses, tenements and messuages built on a former garden plot, grouped around a common yard. Twenty-two households lived there in 1637 and 30 in 1666.[56] The population of the parish of St Botolph Aldgate in which the alley lay grew from c. 1,500 in 1540 to at least 11,000 by 1640 and c. 21,000 by 1710. The displaced populations of the Fire may have further increased demand for and density of accommodation in the inner suburbs, at least in the short term.

Plague is the only epidemic disease for which detailed study of its topographical distribution can be undertaken, given that plague deaths were separately noted in the Bills of Mortality from the first. As Paul Slack showed in 1985, the worst plague mortality in the city shifted from the centre to the inner suburban parishes in the seventeenth century.[57] Justin Champion, in his study of the 1665 epidemic across the whole metropolis, documented a broadly concentric pattern of plague mortality with the highest rates towards the periphery, and again at the general level, a significant correlation between areas where households averaged fewer hearths and higher crisis mortality rates. However, analysing percentages of households infected and average deaths per infected household, he noted considerable local variation, suggesting that 'locational or spatial qualities' might be more strongly linked than social and economic categories. But the picture is greatly complicated by the propensity of the rich to flee, leaving large houses with perhaps only one or two caretakers, and the relationship of plague mortality to gender and of gender to poverty, so that, while he agreed with Slack on a general 'correlation between wealth, poverty and disease', incidence and severity at household level defy prediction.[58] Studies of plague in specific parishes also suggest that households across the wealth spectrum were affected.[59]

One of Champion's key findings is that the 1665 plague deaths, at least, did not markedly cluster in households; 'the most common experience of mortality was singular rather than multiple'.[60] Was this perhaps surprising phenomenon unique to plague, for whatever combination of epidemiological and social reasons? Were London's other epidemic diseases such as smallpox and typhus, generally thought to be encouraged by close living conditions, less discriminating? The bills do not record the incidence of diseases or causes of death other than plague by parish, and the reporting of causes of death in parish registers is sporadic and often short-term, so it is not clear how far this enquiry can proceed. However, as T. R. Forbes showed, the parish records of St Botolph Aldgate offer data on cause of death from 1583 to 1599, as well as considerable detail on the population of the parish, and though he did not analyse this by location or household, this is one of the objectives of the research project 'Life in the Suburbs: Health, Domesticity and Status in Early Modern London'.[61] It may also be feasible to use the bills to identify years with high mortality from a particular disease (such as smallpox in 1652, or fever in 1686)[62] and then to seek to pinpoint this unusual mortality at a more local level, though year-to-year instability is itself a characteristic of the mortality of the period.

2.3.ii   Water, sanitation, and cleanliness

A second aspect of the poorer cheaper housing of the inner suburbs was the lack of amenities such as water and sanitation. The city centre was fairly well supplied with conduits, bosses, and wells or pumps, and the New River Company was piping water to the city centre by the early seventeenth century. The rent lists of the companies providing piped water show, not surprisingly, that they laid their lines to the wealthier areas of the town; piped water was an important consideration in high-class

West End development, but much less is heard about it in poorer areas. The majority of Londoners still collected water from public conduits or watercourses, including the Thames, or bought it from street sellers who obtained it from the same sources.

The extent to which new developments benefited from services and amenities varied; a conscientious landlord might make a real effort to support his or her development with services, but this would not always be worthwhile. Despite the intention to attract high-status tenants, the original licence for the development of Covent Garden made no provision for water supply. Some tenants had access to springs, though these proved unreliable, and others tapped into the City's water pipes leading from Conduit Mead, until in 1633 Bedford and other developer-landowners obtained permission to pipe water from Soho.[63] Shadwell – unusual among eastern suburbs in having a single, somewhat paternalistic developer – had a pumped water supply after 1680, but it is not clear how many individual houses benefited.[64]

The outer suburbs in general may have had comparatively good access to decent water supplies, in the streams and watercourses flowing towards the city, but the inner suburbs had neither, though they may have been served by peripatetic water sellers. The Thames was an important water source for riverside parishes, but was used also for disposing of rubbish; if waterside location was a significant adverse variable for infant mortality rates, as identified by Finlay,[65] this presumably also had implications for child and adult health.

But just as landlords maximised occupation of their properties, they were also inclined to neglect investment in sanitation and waste disposal. A Common Council order of 1570 called for the speedy remedy of ancient alleys 'that [...] have only a common house of easement for the whole alley'. Likewise, where houses had recently been turned to multi-occupation, the owners were required either to reinstate them or to ensure that they were properly divided 'with sufficient [...] houses of office'. Although these were general regulations, the instances singled out for special sanction came from the inner suburbs like Farringdon Without.[66] It was not just private landlords who were at fault: tenants of parish houses belonging to St Botolph Aldgate had to take the vestry to law to force them to provide the tenants with privies, something not guaranteed in their leases; but only four new privies were built for some 13 premises. In general, as both plans and property narratives show, alley neighbours shared privies. In one suburban close, Pheasant Court in Smithfield, three privies served 12 and possibly 14 households.[67] In Three Kings' Alley in Portsoken it appears that 13 houses shared a common well and a common privy in the mid-seventeenth century, and the 13 houses contained 30 households by 1666. We read also of complaints that landlords were recalcitrant about having these emptied or repaired as necessary.[68] In the newer suburbs and the areas where many houses had gardens and yards the situation may have been slightly better.

Not surprisingly, gastric and enteric complaints figure quite largely in the Bills of Mortality. Flux, bloody flux and scouring feature from the first, but were overtaken in the 1650s by 'plague in the guts', soon renamed 'griping in the guts'. An apparent decline in bloody flux was more than matched by high levels of 'griping in the guts'. Probably both of these were dysentery or dysentery-like attacks; Graunt speaks of '*Dysentery*, called by some *The Plague in the Guts*', while the physician Sydenham writes of '*Dysenterick Fever*' in the 1670s being succeeded by bloody flux and of both as characterised or accompanied by gripes.[69] It has been suggested that 'griping in the guts' was a form of infantile diarrhoea, as its eighteenth-century decline is matched by an increase in deaths from convulsions, a disease of infants,[70] but the identification seems less probable for the seventeenth century. Adult deaths were attributed to this cause in St Helen's parish in the 1650s,[71] and Sydenham discusses gripes and bloody flux largely as an adult ailment, noting that 'though this disease is very often deadly in the Adult, and especially in old people, yet 'tis very gentle in Children', though he did also give instructions how to treat 'infants afflicted with this disease'.[72] At any rate, these complaints seem to have been particularly severe in the third quarter of the seventeenth century, accounting for over 12 per cent of non-plague deaths in the 1660s to 1680s, declining to about 9 per cent by 1700.[73] As with epidemic diseases other than plague, however, precise location of mortality is difficult, and if a high level of gastric and enteric mortality seems a predictable concomitant of poor water supply and sanitation, the surprise is rather in the apparent improvement in the later seventeenth century despite London's continued growth.

2.3.iii Warmth and shelter

The final query about housing quality and health is whether the former contributed to the pulmonary and respiratory complaints that seem to have been so common. John Evelyn believed that '*Catharrs, Phthisicks, Coughs* and *Consumptions* rage more in this one City than in the whole Earth besides', exclaiming 'is there under Heaven such *Coughing* and

Snuffing to be heard, as in the *London* Churches and Assemlies of People, where the Barking and the Spitting is uncessant and most importunate?'.[74] This was not merely a nuisance or inconvenience: 'almost one half of them who perish in *London, dye of Phthisical* and *pulmonic* distempers; [...] the *Inhabitants* are never free from *Coughs* and importunate *Rheumatisms,* spitting of *Impostumated* and corrupt matter', while the ulceration of the lungs is 'a mischief so incurable, that it carries away multitudes by Languishing and deep *Consumptions, as the Bills of Mortality* do Weekly inform us'.[75]

Evelyn here assimilates consumption with other pulmonary complaints, and 'Consumption' in the bills was probably not confined to modern pulmonary tuberculosis, but covered a range of diseases characterised by cough, including coughing blood, inflammation or ulceration of the lungs, and wasting of the body. Graunt notes of *Tyssick* (phthisis) 'that it is probable the same is entred as *Cough,* or *Consumption'*, and other contemporaries wrote of 'Phthisicks and consumptions of the lungs'.[76] 'Consumption' may also have embraced pneumonia, 'Peripneumonia' and pleurisy, all identified by contemporaries as inflammations of the lungs,[77] though there is also a separate bills category of pleurisy. Sydenham wrote of the 'Epidemick Coughs, with a Pleurisie, and Peripneumonia coming upon them' that flourished in 1675, and considered that 'Peripneumonia [...] differs only from a Pleurisie in degree'.[78] Consumption's still broader interpretation as wasting, the final symptom of many diseases, but especially venereal disease, may to some extent have confused the issue and the statistics,[79] but 'Consumption, occasioned by distemper'd Lungs [...] is the most deplorable, and what we most frequently meet with, our English people being in a very particular manner subject to it'.[80]

Respiratory ailments were certainly a major cause of death in early modern London. In Graunt's tabulation of the mid-seventeenth century Bills of Mortality, they produced around 25 per cent of deaths across London in non-plague years. 'Consumption' alone caused 15–20 per cent of non-plague deaths between 1629 and 1636, rising to nearly 24 per cent in the later 1640s and 1650s.[81] At the time Evelyn was presumably writing *Fumifugium,* in 1660, they were still over 20 per cent. After 1665, while absolute numbers of deaths attributed to consumption remained high, they declined as a proportion of all deaths, though this was partly the result of increasing infant and smallpox mortality.[82]

Evelyn, as is well known, attributed all the respiratory and pulmonary diseases and their effects to coal smoke, especially from industrial processes.[83] In many cases the 'smoaks and stinkes' of London would have exacerbated symptoms, rather than causing or spreading specific disease, but other environmental factors contributed, in addition to the overcrowding already noted. The poorer houses we have been considering suffered from poor maintenance, poor surface drainage, and fewer hearths per house or household, even if the occupants could afford fuel to burn. Rooms were low and small, as plans and surveys indicate and as epithets like 'close', 'heaped', 'smothered' imply. Ventilation was probably poor too, since glazed windows were rarely noted and the houses opened onto narrow courts and alleys. Cold, damp rooms, inadequate heating and ventilation, when combined with close crowding, would have provided a good breeding ground for respiratory diseases, whether described as cold, cough, 'tissick', consumption, or pleurisy. The view that 'consumption is induced by the foul air of houses' was firmly held up to the nineteenth century,[84] though it should also be noted that both tuberculosis and respiratory infections have a positive correlation with poor nutritional status, and hence by implication poverty more generally.[85]

No direct correlation of deaths from respiratory ailments and housing quality or topography can be made for most of the period, though it may be significant that the proportion of consumption deaths declined after 1660, when causes of death in seven 'distant' parishes (St Margaret Westminster, Islington, Hackney, and Stepney in Middlesex, and Rotherhithe, Newington and Lambeth in Surrey) were added to the whole. Other things being equal, this suggests that respiratory ailments were worse in the crowded inner suburbs, and that greater distance from the centre was beneficial to health, as indeed contemporaries believed.[86] Graunt contrasted '*Hackney, Newington,* and other Country-Parishes' with 'the most *Smoaky,* and *Stinking* parts of the City' and argued that 'the *Smoaks, Stinks,* and close *Air* [of London] are less healthfull than that of the Country'.[87] But one of the few such comparisons we can make indicates that consumption was probably a much greater cause of death in late sixteenth-century St Botolph Aldgate, a rapidly-growing but poor inner suburban parish, than these seventeenth-century city-wide averages. The parish registers give 'con' as the cause of some 35 per cent of non-plague deaths between 1583 and 1599. Even if 'con' conflates convulsions (a cause of infant death) with consumption, the latter still appears to have caused a disproportionately high percentage of deaths.[88] It was also to some extent age-related, accounting for an increasing proportion of deaths in each age group in the parish from the 20s to the 70s.[89] Housing and living conditions of the kind we have been considering, therefore, likely made a major contribution to the incidence and spread of one of the most significant causes of death of the period.

2.4   Conclusion

As noted at the beginning of this paper, close correlation of housing and health outcomes in early modern London is difficult, but there are certainly connections to be inferred. Andrew Appleby found very limited correlation between the nutritional status of London's populace (inferred from the varying price of bread) and the chronological incidence of disease,[90] but this does not mean that the health of the poor was not adversely affected by the physical conditions in which they lived. Environmental conditions in London worsened over the sixteenth to seventeenth centuries, as the capital's population grew tenfold and the majority of new housing was poorer, more crowded, and less well-serviced than before. Royal and civic proclamations against subdivision and inmating seem to have been ineffective. Contemporaries believed that overcrowding spread epidemic disease, even if studies of plague suggest that the picture at the very local level may be complicated, and analysis of other causes of death highlights two important categories – gastric and respiratory complaints – to which environmental and specifically housing-related conditions may well have made a significant contribution.

Notes

1. J. Graunt (1662). *Natural and political observations mentioned in a following index, and made upon the bills of mortality. By John Graunt, citizen of London. With reference to the government, religion, trade, growth, ayre, diseases, and the several changes of the said city*. London, printed by Tho: Roycroft, for John Martin, James Allestry, and Tho: Dicas, at the sign of the Bell in St. Paul's Church-yard (1st edition, ESTC R13975); ibid. (1662) (2nd edition ESTC R12046); ibid., printed by John Martyn, and James Allestry, printers to the Royal Society, and are to be sold at the sign of the Bell in St Pauls Church-yard (3rd edition, 1665: ESTC R11688); ibid., printed for Samuel Speed at the Rainbow in Fleet-street (4th edition, 1665) ESTC R9023); ibid., Oxford, printed by William Hall, for John Martyn, and James Allestry, printers to the Royal Society (1665: ESTC R11654); ibid., London, printed by John Martyn, printer to the Royal Society, at the sign of the Bell in St. Paul's Church-yard (5th edition, 1676). Bibliographical details and all quotations from sixteenth- and seventeenth-century printed sources are from Early English Books Online, unless otherwise stated.
2. Graunt (1662). *Natural and political observations*, p. 69.

3. Ibid., p. 45.
4. R. Finlay (1981). *Population and Metropolis, the Demography of London 1580–1640* (Cambridge: Cambridge University Press); P. Slack (1985) *The Impact of Plague in Tudor and Stuart England* (Oxford: Oxford University Press); J. Champion (1995). *London's Dreaded Visitation: The Social Geography of the Great Plague in 1665* (Historical Geography Research Series 31); J. Landers (1993). *Death and the Metropolis. Studies in the Demographic History of London, 1670–1830* (Cambridge: Cambridge University Press).
5. Champion, *London's Dreaded Visitation*, pp. 50–1.
6. See http://www.history.ac.uk/cmh/pip/index.html for details; see also http://www.geog.cam.ac.uk/research/projects/heahlondon/, [last] accessed 8 August 2011.
7. I have also benefited from stimulating discussions with Steve Cornish, currently working on a PhD thesis applying modern theories of building pathology to the question of building-related illnesses in early modern London.
8. For discussion of London housing types, see J. Schofield (1984). *The Building of London from the Conquest to the Great Fire* (British Museum: London); idem, (1995). *Medieval London Houses* (New Haven and London: Yale University Press). For an overview of housing and urban change, see V. Harding (2001). 'City, capital and metropolis: the changing shape of seventeenth-century London', in J. F. Merritt (ed.). *Imagining Early Modern London: Perceptions and Portrayals of the City from Stow to Strype, 1598–1720* (Cambridge: Cambridge University Press), pp. 117–43; idem (2007). 'Families and housing in seventeenth-century London', *Parergo*, 24(2), 115–38.
9. For an account of the geographical division of London in the bills, see V. Harding (1990). 'The population of early modern London: a review of the published evidence', *London Journal*, 15, 111–28.
10. Records of these kinds constitute the basis for the historical reconstructions of houses in the five Cheapside parishes: D. Keene and V. Harding (1987). *Historical Gazetteer of London before the Great Fire*, available online via British History Online: http://www.british-history.ac.uk/source.aspx?pubid=8, [last] accessed 8 August 2011. For an example of drawn plans and surveys, see J. Schofield (ed.) (1987). *The London Surveys of Ralph Treswell* (London: London Topographical Society).
11. J. S. Loengard (ed.) (1989). *London Viewers and their Certificates, 1508–1558: Certificates of the sworn viewers of the City of London*, available online via *British History Online*: http://www.british-history.ac.uk/source.aspx?pubid=158, [last] accessed 8 August 2011.
12. For houses of this kind, see Keene and Harding, *Historical Gazetteer*.
13. Cf. e.g., M. J. Power (1986). 'The social topography of Restoration London', in A. L. Beier and R. Finlay (eds). *London 1500–1700, the Making of the Metropolis* (London and New York: Longman), pp. 199–223.
14. See T. C. Dale (ed.) (1938). *The Inhabitants of London in 1638*, available online via *British History Online*: http://www.british-history.ac.uk/source.aspx?pubid=176, [last] accessed 8 August 2011; Finlay, *Population and Metropolis*, Chapter 4; E. Jones (1980). 'London in the early seventeenth century: an ecological approach', *London Journal*, 6, pp. 123–34.
15. Graunt, *Natural and political observations*, pp. 53–4.

16. For an account of the development of alleys and closes, see Harding, 'Families and housing'.
17. Power, 'The social topography of Restoration London', fig. 4, p. 207.
18. M. J. Power (1978). 'The east and west in early-modern London', in E. W. Ives, R. J. Knecht and J. J. Scarisbrick (ed.). *Wealth and Power in Tudor England* (London: Athlone Press), p. 170.
19. A. L. Beier (1986). 'Engine of manufacture: the trades of London', in A. L. Beier and R. Finlay (eds). *London 1500–1700, the Making of the Metropolis* (London and New York: Longman), pp. 141–67.
20. See N. G. Brett-James (1935). *The Growth of Stuart London* (London: George Allen & Unwin); L. Stone (1980). 'Residential development of the west end of London in the 17th century', in B. C. Malament (ed.). *After the Reformation* (Manchester: Manchester University Press), pp. 167–212.
21. F. H. W. Sheppard (ed.) (1970). *Survey of London*, vol. 36, *Covent Garden*, available online via *British History Online*: http://www.british-history.ac.uk/source.aspx?pubid=362, [last] accessed 8 August 2011.
22. M. J. Power (1978). 'Shadwell: the development of a London suburban community in the seventeenth century', *London Journal*, 4, 1, 29–46; idem, 'East and west'; idem, 'East London housing in the 17th century', in P. Clark and P. Slack (eds) (1972). *Crisis and order in English towns 1500–1700* (Cambridge: Cambridge University Press), pp. 237–62.
23. Power, 'East and west', p. 170.
24. Power, 'The social topography of Restoration London', fig. 3, p. 203.
25. Power, 'East and west', p. 170; idem, 'The social topography of Restoration London', Table 24, p. 205.
26. J. Strype (1720). *A survey of the cities of London and Westminster* (London), bk 4, ch. 2, pp. 39–43: online edition, Humanities Research Institute, University of Sheffield, http://www.hrionline.ac.uk/strype/, [last] accessed 8 August 2011.
27. Strype, *A survey of the cities of London and Westminster*, bk 4, ch. 2, p. 48.
28. Jones, 'London in the early seventeenth century', pp. 123–34.
29. Power, 'The social topography of Restoration London', pp. 202–6.
30. C. Spence (2000). *London in the 1690s. A social atlas*, London: London Topographical Society, pp. 66–75.
31. Based on figures and assumptions discussed in Harding, 'Population of London', with further calculations for the population of the distant parishes.
32. T. F. Reddaway (1940). *The Rebuilding of London after the Great Fire* (London: Jonathan Cape).
33. See especially Graunt, *Natural and political observations* (1662).
34. Graunt, *Natural and political observations* (1662); T. Birch (ed.) (1759). *A Collection of the Yearly Bills of Mortality, from 1657 to 1758 inclusive …* (London), available online via ECCO (Eighteenth-Century Collections Online). *A Collection* also includes a reprint of Graunt's 'Observations' and tables. Appleby tabulates the bills' figures for epidemic mortality 1629–1750: A. B. Appleby (1975). 'Nutrition and disease: the case of London, 1550–1750', *Journal of Interdisciplinary History*, 6, 1–22.
35. Finlay, *Population and Metropolis*, Table 5.12, p. 102; Graunt, *Natural and political observations* (1662), p. 14.

36. The research project People in Place contrasted infant mortality in Clerken-well and Cheapside (http://www.history.ac.uk/cmh/pip/pip.html#env), [last] accessed 8 August 2011, while Finlay contrasts inner-city inland parishes with two intramural riverside parishes: *Population and metropolis*, pp. 101–6.
37. http://www.history.ac.uk/cmh/pip/pip.html#env.
38. Slack, *The Impact of Plague*, pp. 145–72, esp. Fig. 6.1, p. 146; Champion, *London's Dreaded Visitation*.
39. Graunt, *Natural and political observations* (1662). p. 36: 'It may be also noted, that many times other *Pestilential* Diseases, as *Purple-Feavers, Small-Pox*, &c. do forerun the *Plague* a Year, two or three'.
40. Graunt, *Natural and political observations* (1662). 'Table of casualties'.
41. Appleby, 'Nutrition and disease', pp. 20–1.
42. Calculated from data in Birch, *A Collection of the Yearly Bills*.
43. Calculated from data in Birch, *A Collection of the Yearly Bills*.
44. Graunt, *Natural and political observations* (1662), p. 16.
45. Ibid., p. 18.
46. Landers, *Death in the Metropolis*, pp. 301–15.
47. Proclamation of 1580, quoted in Strype, *A survey of the cities of London and Westminster*, bk 5, ch. 31, p. 436.
48. Ibid., bk 4, ch. 1, p. 19.
49. E.g., Strype, *A survey of the cities of London and Westminster*, bk 2, ch. 2, p. 16; ibid., bk 2, ch. 6, p. 96; ibid., bk 3, ch. 6, p. 93; ibid., bk 3, ch. 12, pp. 256, 278; ibid., bk 6, ch. 6, p. 87.
50. Based on population estimates (see Harding, 'Population of London') and parish acreages (Ordnance Survey, 1881, listed in Finlay, *Population and Metropolis*, pp. 170–2).
51. Schofield, *London Surveys of Ralph Treswell*, pp. 132–4.
52. The National Archives (TNA): Public Record Office (PRO), SP 16/359, original returns of divided houses. T. C. Dale, 'Returns of divided houses in the city of London (May) 1637' (typescript transcription of the returns, 1937) in Guildhall Library Printed books, SL 06.2.
53. Dale, 'Returns', p. 147.
54. Ibid., pp. 136–7.
55. Ibid., pp. 122–35.
56. Centre for Metropolitan History, Institute of Historical Research: draft 'Aldgate Gazetteer', Three Kings Alley 43/7/3.
57. Slack, *Impact of Plague*, chapter 6.
58. Champion, *London's Dreaded Visitation*, esp. pp. 42–52, 62–80.
59. Unpublished paper by Philip Baker to 'Housing environments and health' symposium.
60. Champion, *London's Dreaded Visitation*, p. 70.
61. T. R. Forbes (1971). *Chronicle from Aldgate: Life and Death in Shakespeare's London* (New Haven and London, Yale University Press); http://www.geog.cam.ac.uk/research/projects/lits/, [last] accessed 8 August 2011.
62. Appleby, 'Nutrition and disease', pp. 20–1.
63. Sheppard, *Survey of London*, pp. 25–34.
64. Power, 'Shadwell'.
65. Finlay, *Population and Metropolis*, p. 103.

66. London Metropolitan Archive, City of London Records Office, Journal 19 f. 255, 13 June 1570.
67. Schofield, *London Surveys of Ralph Treswell*, pp. 132–4.
68. Centre for Metropolitan History, Institute of Historical Research: draft 'Aldgate Gazetteer', Three Kings Alley 43/7/3.
69. Data from Birch, *A Collection of the Yearly Bills*; cf. Graunt, *Natural and political observations* (1662). 'Table of casualties'; T. Sydenham (1696). *The Whole Works of that Excellent Practical Physitian Dr Thomas Sydenham*, translated by John Pechy, M. D. (London: printed for Richard Wellington, at the Lute, and Edward Castle, at the Angel, in St Paul's Church-yard), pp. 127–44.
70. C. Creighton (1891, 1894). *History of Epidemics in Britain*, cited and discussed in Landers, *Death and the Metropolis*, pp. 94–6.
71. W. B. Bannerman (ed.) (1904). *The Registers of St Helen, Bishopsgate* (Harleian Society 31), pp. 302, 304.
72. Sydenham, *The Whole Works*, pp. 134, 139.
73. Calculations from data in Birch, *A Collection of the Yearly Bills*.
74. J. Evelyn (1661). *Fumifugium, or the Inconveniencie of the Aer and Smoak of London dissipated* (London: printed by W. Godbid for Gabriel Bedel, and Thomas Collins, and are to be sold at their shop at the Middle Temple gate neer Temple-Bar), p. 5, 10.
75. Evelyn, *Fumifugium*, p. 13.
76. Graunt, *Natural and political observations* (1662), p. 29; *Oxford English Dictionary* (*OED*), 17th-century examples under consumption, phthisis, phthysic. Cf. B. Marten (1720). *A new theory of consumptions: more especially of a phthisis, or consumption of the lungs* (London).
77. *OED*, 17th-century examples under consumption, pneumonia, peripneumonia and pleurisy.
78. Sydenham, *The Whole Works*, pp. 197–205, 221.
79. *OED*, 'consumption', citing R. Wittie, tr. J. Primrose (1651). *Pop. Errors*, II. 88: 'They doe not distinguish the true consumption from other diseases, but call every wasting of the body, a consumption'.
80. Marten, *A New Theory of Consumptions*, pp. 2–3.
81. Graunt, *Natural and political observations* (1662), 'Table of casualties'.
82. Data from Birch, *A Collection of the Yearly Bills*.
83. Evelyn, *Fumifugium*, p. 5 and passim.
84. *OED*, 'consumption', citing F. Nightingale, *Notes on Nursing* (1861), p. 26.
85. Landers, *Death and the Metropolis*, p. 30.
86. Data from Birch, *A Collection of the Yearly Bills*.
87. Graunt, *Natural and political observations* (1662), pp. 45–6.
88. Forbes, *Chronicle from Aldgate*, pp. 100–2 (1,101 of 3,086 non-plague deaths).
89. Forbes, *Chronicle from Aldgate*, Table 7, p. 103. Percentages of all deaths by age-group cannot be calculated from Forbes's data, but 'con' accounted for 94 deaths of people in their 20s, while other common causes (excluding 'long sick' and 'aged') accounted for 145 more; 'con' and other causes of death for those in their 60s were 112 and 55 respectively, and for those in their 70s, 67 and 9.
90. Appleby, 'Nutrition and disease', 1–22.

3
Environment and Disease in Ireland

Christopher Hamlin

3.1 Introduction

This chapter uses the case of famine-era Ireland to explore the challenge of the natural – or better, the biogeochemical – domain for historical scholarship generally, and, secondarily to explore tensions between medical and environmental history.

How and how much to bring nature into the writing of history? In principle, bringing nature into narrative – incorporating the meteorological, the mycological, and all between – would seem uncontroversial. It usually is, and yet a heavy enough dusting of natural background can naturalise an entire account, overthrowing conventions of epistemology, explanation and assessment. Medical and environmental historians are the primary gatekeeper-mediators of that potential. Of course, like other historians, they write about knowledges, institutions and apprehensions, and about contingent human actions. Both fields comprehend multiple methods, questions and agendas, so generalisations are tricky, but they share a claim to speak for the biotic. Medical history subsumes the biotic status of humans, while environmental historians explore not only apprehensions of exteriority, but droughts and frosts, exhausted soils and fisheries, and pervasive toxins. There are tensions – the one focuses more on the biology of single bodies, the other on communities and populations – still each domain assumes the other.

For each, the biological is likelier to be text than context. If these fields matter at all, it is because the biotic matters – it limits, enables, determines. But whenever determinism threatens to displace contingency, medical and environmental historians are apt to find their work well walled-off. A biotic story may co-exist with the human (non-biotic) story: we may know that both stories pertain to the same domain of events and

45

yet be unable to negotiate common tenancy.[1] The problems are not new. The *Annalistes* sought to foreground soils and infections. Yet, while many admire Braudel, few follow: to a profession preoccupied with human-made change and tied to a document-dominated conception of human action, *longue durée* often meant 'nothing happened'.[2] To the degree that environmental and medical factors manifest only as alternatives to human doings, mainstreaming will remain problematic. But that very tension has been central in the history of the Irish famine.

I first explore the valences of nature and culture through contrasting representations of famine and of epidemic disease. I turn then to 'fever' and 'bogs'. Both were, in the terms of recent scholarship, nature-culture 'hybrids', but more importantly they were loci of translation. By means of 'fever', one form of culture – obligation to the hungry – became 'nature', though the translation ultimately collapsed in an intra-professional dispute. 'Bogs,' on the other hand, were nature made into culture, the locus of an intoxicating vision of healthy and sustainable peasant empires. They were the fetish that allowed continuation of structural contradictions.

3.2   Representations: historians

In famine Ireland, between 1845 to 1852, a population that had rapidly risen rapidly fell from 8.2 to 6.6 million. Incidental disasters have been of little interest if they are merely acts of nature. Yet, the key issue in the Irish famine has been what Austin Bourke calls 'malintent'.[3] Mainly this is directed towards responders, whose inaction comes close to rendering them perpetrators, a view the meteorologist-historian Bourke rejects. Yet, however much they are products of policy, famines are also matters of winds and soils, and perhaps moulds and *Rickettsiae*, which interact with human bodies that need nutrients to flourish, act and document their existence. They also reflect structures: the problem is not simply of an unusually large number of individual humans who happen to be hungry at the same time, but of a society prone to the recurrence of mass hunger episodes.

Many accounts struggle with the nature-culture border. Is the problem simply the potato blight in late summer 1845, following prolonged wet weather (nature)? Or is it inadequate relief and continued exports – the famous grain-laden ships leaving Cork harbour while people starved onshore (culture, genocide)?

Some seek integration in economics. Cormac Ó Gráda notes that the crop lost to blight in the late 1840s far exceeded that exported: halting exports would have dented the deficit by one seventh at best.[4] The exports also marked an incipient but underdeveloped wage-market economy: more cash would have rendered Ireland more resistant to a short-term deficit in a staple. And export income helped buy food. Yet an analysis embedded in a naturalised economics hides spatial-political issues. The ostensibly 'united' Kingdom of Great Britain and Ireland was not: its regions differed profoundly in recognition of a right to the means of survival. Famines, Sen makes clear, are about distribution as well as production; for many it is distributive choice that dominates famine history.[5]

It also hides ethical issues. Behind a naturalised economics lie ecological analyses. Some contemporaries appealed to a concept of carrying capacity to unite production (nature?) with consumption (culture?): pre-famine Ireland, they claimed, was overpopulated. Malthusian questions are a part of environmental history, notes Donald Worster: we do and should ask 'How many humans can the biosphere support?'[6] But one may ask too, how is the biosphere in question defined and ordered? In famine-era Ireland, attempts to augment prosperity by investing in market (export) agriculture, thereby to balance production with consumption, incidentally involved lowering consumption: estates were to be cleared, human numbers were to shrink so that prosperity (capital, and possibly numbers?) might grow.

As a descriptive analytic, I shall use Worster's concept of agroecosystems. The approach integrates all aspects of a productive system, not only soils, climate and cultivars, but also tools, techniques, and institutions.[7] By analogy it invites us to think systemically about pathology. But such an analytic cannot alone sort out the tangle of cause, blame and responsibility, which have been the central problems of famine narration. By 'cause' I mean an accounting that an epidemiologist of any stripe might give, by 'blame,' a retrospective moral assessment of motive and obligation, and by 'responsibility,' a praxis of response. Often they are conflated. In making mycology exculpatory, cause is made to serve as blame.

Most writers on the Irish famine have followed Sen: ultimately famine is failed obligation and culpable act, more a matter of killing than of dying. Yet that conclusion involves the privileging of certain levels of accounting. Behind it, 'environment' continues to lurk, for whatever else it was, this famine represented changes – in weather and in the corresponding prominence of a blight-causing mould. So too, the presence of epidemic disease complicates. The vast majority of 'famine' deaths were not from hunger but from typhus, relapsing fever, dysentery, or cholera. That total, noted the pioneering famine historian Cecil Woodham-Smith, 'will never be known, but probably about ten times more died of disease

than of starvation'. Surely this is 'nature' – better even to call the event
an epidemic? ('Involuntarily, we found ourselves regarding this hunger
as we should an epidemic, looking upon starvation as a disease', wrote
the Bradford Quaker W. E. Forster, after a tour of the west of Ireland in
late 1846.)[8] Generally – with the exception of the cholera (part of a world
pandemic) – the presumption is that famine and disease were more than
coincidental, but the links are unclear. And yet, Ireland's earlier 'great
famine', in 1816–19, likewise the outcome of two years of harvest failures,
had been managed as a fever epidemic. While the 'great famine' of late
1840s has had world significance as an exemplar of colonial indifference,
the earlier event is virtually invisible even within Irish history.[9]

To translate genocide as disease – even disease with some causal tie to
famine – is to risk redefining an event and moving victims even further
from culture into nature. While some might ask in what Orwellian world
it matters whether people starve or die of disease before they can starve,
it does. Voyages of lice and Rickettsiae lack the valences of grain-laden
ships; they have no place for contingency and agency, and thus enor-
mity. Even if we allow that epidemics were tied to scarcity, it can matter
whether the link is incidental (famine displacement spreads contagion)
or essential (famine weakens resistance).[10] Hence disease-as-cause-of-
death has often been a fact to be gotten past – couldn't the stricken have
held on longer to register their victimisation unambiguously?

This problem of accommodating 'nature' is not mainly one of inclusion but of integration and explanation. It may seem at first sight that
'culture' – site of choice and change – is accusatory while 'nature' is
exculpatory: it represents 'inevitability,' master source for apologists.
Further, to render 'subjects' in terms of nature has been seen to under-
cut their humanness, understood as agency. And yet any serious claim
of agency rests on a claim of capacity, which translates readily into
health.[11] The 'nature-culture' hybrids now in vogue are less resolution
than prosecution rebuttal. In them, 'nature' is human creation, an apol-
ogist's tool. In the long run, 'nature-culture' does the prosecution no
good either, for its case relies ultimately on a force of nature. Injustices
are measured in currencies of nature; in bodies (not 'bodies') that thrive
or wither – and suffer.[12] Taken to its end, it would threaten body as well
as embodiment and give us a history of medicine without disease, but
also without the bodies that may suffer.[13]

Hunger has been especially hard to register in terms other than
height, weight or (presumed) caloric intake.[14] The very hungry or weak
may be unable even to speak, much less make historical documents.
Our traces of their wordless suffering come from observers who are

more or less horrified, outraged or moved to sympathy, and struggle to express coldness, tiredness, weakness, pain or hopelessness to which they have no direct access.[15] In histories of the Irish famine, hungry bodies move back and forth across the border separating culture from nature. Far from being 'hunger artists', in Ellman's terms, they are hunger canvases.[16] Their cultural place may be plain, but the impossibility of registering their modes of disability has meant that they are embodied only as the not-yet-dead. Once dead, again unambiguously biological, they are countable and available for enlistment in the stories others might want to tell. Remarkably, this is not the case with other diseases – in consumption or cancer, mania or polio there are options for bringing biology into one's self-accounting.

Nor does hunger fit readily into disease models. Notwithstanding its prodigious impact in human history, hunger has gotten little attention from medical historians. The silent vanishing of inanition that barely ruffles surrounding society requires no diagnostic validation, no complicated therapeutic considerations. The immediate causes of starvation are plain, so too the solutions. Famine has no clear point of entry in a medical marketplace; not only does it not necessarily involve the recognised physician, it is unlikely to involve any sort of medical practitioner, since the very existence of famine is often evidence of the inability to act as consumer in any marketplaces whatsoever, not only the medical. Hence, as Davis suggests, there has usually been no predicate for sentences that begin with diagnoses of chronic malnutrition.[17]

3.3 Representations: contemporaries

Culture-nature tensions have been prominent in recent historical writing on events in Ireland in the late 1840s, but medical, environmental and political accountings of these events have not always been antithetical. Early nineteenth-century Ireland was the site of remarkable public discussions on determinants of poverty and disease, which took place within a loose medical-police framework. Somehow, the reformism sweeping England would in time be applied to Ireland; the success of Catholic emancipation, and the acuity of series of lords lieutenants and chief secretaries seemed for a time to open a variety of possibilities. A frame of 'nature' versus 'culture', in which natural events lay outside government accountability only began to harden in the late 1830s in conjunction with conflict over the new Irish poor law and the complexities of land reform, but also in response to the dashed hopes for political reform.

Even before the famine, Ireland was being represented in pathological terms. Disembarking in Dublin in 1844, the American evangelical Asneath Nicholson anticipated 'the dark curtain of desolation and death' of 'bleeding, dying' Ireland. One of the first she met on the street declared, 'Oh, the country's dyin', it's starvin', it's kilt'.[18] This was a good year. The massive inquiries of the 1834–36 Royal Commission on the State of the Irish Poor were likewise undertaken in good years, but disclose regular hunger among a quarter of the population during early summer.

Accordingly, the great famine of the later 1840s had been dichotomised *before* it happened. One side appealed to nature. Frustrated with the persistent poverty and resistance to economic rationalisation, many evangelicals and political economists, including the government's chief famine policy maker, Charles Trevelyan, had given up hope of reform, and anticipated a natural revolution. Famine or deadly epidemic was the storm that must finally break God's redemptive rule through natural law. It would fix what political economy-flouting humans had ruined; it would end the Malthusian misery.[19] Among those who endorsed the view – even well after the famine – was the novelist Anthony Trollope, who helped readers of *Castle Richmond* to see the famine as divine mercy. Post-famine Irish labourers now made good wages: supply and demand of labour had equilibrated.[20] If most now see such views as misanthropic, contemporary concerns about world population growth strike a similar tone. Things cannot go on as they have been; sooner or later something must give; some natural but ultimately benign process will kick in when things become too crowded.

On the other side was a rendering in terms of culture and power. Famine was moral and practical error; it reflected Ireland's powerlessness, imperial incompetence, and British (Anglo-Saxon) contempt. While there had been numerous expressions of outrage on all these grounds during the late 1840s, only a decade later did the famine begin to exemplify England's infamy. In *The Last Conquest of Ireland (Perhaps)* (1858), the Young Irelander John Mitchel presented Ireland as the victim not of blight but of political economy. Rightly recognising its intimate association with the evangelical movement (exemplified by Thomas Chalmers' Malthusianism), Mitchel represented political economy as the 'creed and gospel of England' and sneered at its pretence of 'saving doctrines'.[21] The evangelicals' portrayal of impending demographic end times was instead the stability of a viable traditional culture in its place. Any vulnerability was from exogenous *political* events – like theft of its food by foreign rulers. In this timeless and classless Irish authenticity, agroecological, social, demographic or epidemiological factors are rendered as cultural attributes, in need neither of explanation nor of justification. Crowded cabins are sociability; begging – including in the many preceding episodes of hunger and fever – are forms of hospitality and charity; conflict is resistance; and proposed technical or economic reforms are often colonial ideology. The views diverge most sharply in their assessment of the potato: worst of addictive drugs or life-giver to the many. However well his romantic recasting of pre-famine society distinguished perpetrators from victims, Mitchel failed to offer a coherent agroecological critique. Though, as J. S. Donnelly as shown, the famine did not quickly acquire its current prominence in the chronicling of the Irish past, when finally it did, after 1950, it was Mitchel's version that prevailed.[22]

Mitchel recognised the dichotomy between famine and disease I noted earlier. He mocked the Census figure of only 516 starvation deaths during the famine years, and highlighted those that were the verdicts of coroner's inquests, particularly where juries fingered the Prime Minister, Lord John Russell, as culpable. The relevant context was criminal jurisprudence not medicine, because blame and responsibility, and not merely cause, were at stake. He recognised the need to enlist disease, writing of 'hunger, and its sure attendant, Typhus', and referring to deaths 'either of mere hunger, or of typhus fever caused by hunger'. And, finally, he represented these deaths as the execution of public policy. The land rationalisers dictated desettlement; to effect this, natural means of starvation were quicker, surer and cleaner than cumbersome eviction. Even when he contrasted England and Ireland in terms of disease metaphors, Mitchel focused on matters alimentary: 'As dyspepsia creeps into England, dysentery ravages Ireland'.[23]

Though one side appeals to dynamism and the other to stasis, both the catharsis and the cultural genocide account obscure any substantive contingency or choice in environmental and in health policy. Neither gets us close to the experience of suffering from hunger-disease or to efforts to respond to it. However artfully each side enlists disease, neither brings us closer to recognising a multi-dimensional public health problem in which medicine would have been the obvious locus of professional response.

3.4 Representations: the medical profession

That picture of neglect is confirmed – in histories of Irish medicine which focus on the great Dublin clinicians, Robert Graves and William

Stokes, who develop a scientific medicine that is only incidentally Irish – by the image of dear, dirty Dublin, where public health has little purchase, and by a mythos of an amedicate rural Ireland, whose colonial peasantry live and die in their peaceable poverty.[24]

In fact, before 1837, medicine had been a key site of response to poverty. Pre-famine representations of Ireland as desperately ill did translate into medical responses. As infirmary or fever-hospital staff, many of Ireland's doctors were part of what was, *de facto*, a system of social medicine, which one historian has described as 'lavish,' by contemporary standards.[25] Yet, though the profession slowly mobilised to deal with fever in 1846–47, the famine did not register.[26] When *The Dublin Medical Press* acknowledged a famine in mid-1847, it was to grouse that doctors were not getting their share of relief funds and to grieve their martyrdom by the masses of infection-spreading poor.[27] One may well ask what happened between, and how an event experienced as failed health acquired an historiography that neglected that.

In part, the answer is squabbles. Intra-professional rivalry led to what might be called 'a lock out of doctors by doctors' and left the profession unable to follow up a widely shared insight that agroecological origins of disease dictated agroecological solutions – medicine and environment must join in practice because they are joined in fact. But unlike England, where the Health of Towns campaigners found in urban sanitation a way beyond class politics, health reform in Ireland could not be confined to a single innocuous technical fix. It could not transcend deeper structural conflict.

To many, including doctors, pre-famine Ireland's systemic health problem was rural rather than urban, and tied to overpopulation – at least overpopulation relative to the means and organisation of production. To the new economists, who were defining their science in terms of scarcity, zero-sum phenomena were paradigmatic; to the medical profession, traditionally apolitical, dedicated to the well-being of individuals, such a situation was deeply problematic: the curing of individuals assumed the availability of family or community resources that were truly surplus and could be supplied without damaging the health of those who supplied them.

In pre-famine Ireland this was not always the case. Moreover, in a Malthusian perspective, each successful cure represented added stress on systems of production and distribution. Pre-famine Irish doctors had wrestled with the most profound questions of social and political medicine, usually on the premise that there were broad bases for social and environmental reform by which Ireland might be healed physically (and, in the process, politically). Like many contemporary health and development-workers, they conceived health holistically, recognising that such matters as aspiration and empowerment affected both population stability and health outcomes. But they, and others, realised too that in practice many elements of nature-culture were zero-sum, arenas for incessant struggle and quiet compromise. Effective political medicine was often best practiced outside of any public gaze, unaccompanied by pamphlets and without rallying cries.

In Ireland, the dominant 'idiom' of conversations about investment in health was 'fever', an ambiguous clinical construction seen by the greatest contemporary Irish physicians, like Groves and Stokes, as a quintessentially Hibernian disease. If this 'fever' was not new to Ireland, it quickly acquired prominence in the years following the repeal of Irish self-governance in 1801. With the founding of a fever hospital in Waterford in 1801, a rapid expansion of Irish medical charities began. Fever hospitals (and dispensaries) supplemented the existing county infirmaries. By the time of the famine there were about 750 public medical institutions.[28]

A key factor in the conversion of this 'fever' from a series of local medical events to general problem of Irish governance was the 1816–19 epidemic, which killed 65,000 of an estimated million and a half cases in a population of 6 million. In 1817, as fever approached Dublin, the young Robert Peel, beginning his career as Irish chief secretary, consulted the Dublin physician Robert Percival. Familiar with European responses, Percival annotated fever treatises for Peel's use.[29] For Peel, both then and again as the first famine prime minister in 1845–6, the fragility of governance was foremost. But, cognisant of the delicate boundary between imperial and local, he did not seek to rule epidemic response. He did make available government hospitals and expedited legislation establishing central and local boards of health.[30]

But he also sent an important signal that medicine had a key role in the functioning of the Irish state. The doctors heard. In a land ruled by coercion acts, they would become *de facto* state agents. Permeating the countryside, they mediated and mitigated conflict over access to the means of life. Presumed to be clear-sighted, disinterested and decisive, they were a comforting adjunct to the army. Indeed, in 1818, the two institutions were joined at the hip. Importuning petitions from the provinces could not be trusted, so Peel relied on army surgeons. Where needed, they could practice fever medicine among the public.[31]

In Dublin, the doctors' agenda was evident in the health subcommittee of the new Mendicity Society. They showed that Dublin's fever was due more to neglect of its own poor than to diseased strangers from the famished countryside. In early 1818 they proposed a comprehensive programme of inspection, hospitalisation, feeding, disinfection, and reclothing. As medical expertise, the proposal was not negotiable, they made clear to the reluctant Society. The fever rate would explode if steps were not taken. They were not, and it did. Their fifth report in September had a 'we told you so' tone.[32]

For the next two decades, as medical charities spread, Irish doctors were conspicuous in thinking big. Manifestoes of medically assisted government creep into their writings. Take for example, an 1818 letter from Whitley Stokes to a Peel staffer. Praising the administration for its hospitals, Stokes urges its acceptance of 'the health of the people' as a duty of 'enlightened' governments. But he stresses secondary benefits. 'If governments are anxious to prevent Idleness, Mendicity, famine and outrage, can they look with indifference on an evil [fever] that directly leads to Idleness Mendicity Famine, and Outrage[?] If they wish for productive industry ... do they hope it from the palsied hand of sickness?' Stokes avoided cannon fodder arguments, but noted that the health of a garrison was tied to that of the surrounding public.[33]

The keystone of their authority was the threat of 'fever'. While Scottish influences are evident, Irish physicians shared no single concept of what 'fever' was. Rather, fever hospital administration made an already broad concept broader. Initially fever hospitals were primarily institutions of isolation: the wretch in hospital was one beggar fewer to exhale on or paw at one, and the donating classes were thus protected. Yet the strategy depended on appealing to sufferers. The urban and rural poor were encouraged to apply to the hospital early, even before the disease had unambiguously declared itself. Accordingly, hospitals had to win their confidence. The hospital experience was to be a relatively attractive one: usually, a patient received a purge on admission, but thereafter treatment was supportive – food and rest for the famished and exhausted.

A post-germ theoretic perspective will focus on the false positives of such a policy. Reports occasionally mention patients who do not have fever, but the greater response was a blurring of boundaries. Effectively, fever became the physiological concomitant of severe deprivation. Almost uniformly, Irish doctors operated with two axioms. First, this 'fever' was a generic entity. Particular symptoms and courses might be temporarily ascendant, but fever varied along multiple axes. What was called 'true typhus' was one type, but rare; most fevers were milder. Second was a doctrine of emergent contagion: high debility could spontaneously generate deadly contagion that would even affect the undebilitated. A contagion might then breed true, generating a particular fever, but that did not imply any concept of disease specificity. It was this adaptability that made fever hospitals so crucial: they were not meant to be general asylums, but since a mild febrile catarrh might always evolve into a malignant (and contagious) typhus, prudence dictated that no one who felt feverish should be turned away.[34] Thus, just when clinicians elsewhere were splitting 'fever' by anatomico- pathological correlation, Irish doctors were lumping. It made sense: hospitalising many mild cases lowered hospital mortality rates, which would delight donors, attract patients, raise reputations and demonstrate the need for hospital expansion.

There were corollaries. One was that there was a great deal of 'fever' in Ireland, much more than in England or Scotland. In a period in which there were roughly 600 cases at the London Fever Hospital, Dublin's Cork Street treated nearly 24,000.[35] Second, most fever was mild among the poor, dangerous to the rich. Class-based case-mortality rates often differ by an order of magnitude in favour of the poor, varying for example from 4 to 40 per cent, and the obvious explanation is that the designation 'fever' was being applied differently.[36] Third, and most important, contagionist explanations did not compete with environmental or social causes. Emphases differed, but blame, responsibility and cause became entangled in ways that mirrored broader issues of Irish politics and which prefigure modern network and risk-factor thinking more than do most nineteenth-century aetiological frameworks.

All these factors reflect the function of fever as a site of negotiation: often, 'fever', was the way in which hunger could be responded to outside the constraints of political economy. If the debilitated poor could engender deadly disease among the rich, their fate was bound to that of the poor. Hence the doctors' message was 'your money or your life'. Thus fever became a weapon of the weak: the beggar or tramp rewarding belated charity with deadly contagion was a powerful moral text. Fever bound regions together, revealed links between town and hinterland. It was centripetal in a centrifugal society.

As magistrates of the sick role, doctors were brokers in survival and justice. They were go-betweens. So much of Irish social relations – the false deference of the poor 'creature' to 'your honour', the exorbitant, but not always paid rents, the constant threat of agrarian rough justice – reflects the pretences of subaltern societies.[37] Not only was there need for acuity in reading the codes, but for backchannels that would lubricate without appearing to violate structures and roles.

Medical care, translated sometimes as medically authorised food and shelter, was such a lubricant. Given the mess of political, confessional and class division, the usual authorities, clergy and gentry, were less able to act. Presumably able to distinguish genuine need from dissembling, doctors were entrusted to evaluate, translate and sanction the demands of the poor. And yet the independence of the professional's judgment was an illusion; mediation was itself a field of biopower. For, often, Irish dispensary doctors were in no position to dictate who deserved care or what that care should be. Both job tenure and personal safety might be at stake if one did not recognise and respect the local power balance.

If many performed these roles quietly, a few thought the centrality of fever gave them authority to assess Irish institutions on medical criteria. They developed what one, Henry Maunsell, called a 'political medicine', a mode for policy analysis, not merely execution.[38]

3.5   'Fevers'

Again, such concerns evolved. An exemplar is the Cork Street physician Richard Grattan (1790–1886), whose radicalisation can be tracked through successive annual reports. In 1815 Grattan was a victim blamer. Dublin was packed with 'crowds of wretched objects, ... [like] beings of an inferior order. ... [incapable of] ... active exertion'.[39] By 1819 causes had become symptoms, or the results of other causes. Reassessment was necessary because the fever hospitals had failed to stop fever.[40] Rants about the indolent might be useful in identifying an abject and needful population, but, if fever were being continually renewed by debilitated contagion-generating bodies, they would not contribute to effective responses. Thus the causes of debility-poverty became a medical matter. In terms of medical geography, Ireland ought to be healthy, in Grattan's view: hence *it* was not the problem, but its history and governance were. He attributed fever to the eighteenth-century penal laws against Catholics and the 1801 abolition of the Irish parliament. The former had left Ireland with an uneducated peasantry, while the latter had deprived Dublin's artisans of a market. Still, Grattan was only diagnosing, focusing only on Dublin, and on political decisions rather than agroecological systems.[41]

In an 1826 pamphlet, *Observations on the Causes and Prevention of Fever and Pauperism in Ireland*, Grattan went further. To the key question of why Ireland had more fever than England, Grattan replied that wages were better there. In turn, that reflected the fact that in England, 'first principles of natural justice' underlay agro-ecological systems.[42] Anticipating Henry George, Grattan proposed a single tax on land.

Grattan was not writing as a proto-nationalist. His attribution of disease to policies and institutions belongs not to blame, but to cause and responsibility. Cultures of unhealth are deeply seated, Grattan recognises; they will not shift quickly. Improving health is environmental in a broadly integrative sense. It involves the use of land and capital given exigencies of population and place.

Others too – Francis Rogan and Henry MacCormac in Belfast, William Pickells in Cork, Francis Barker, John Cheyne, Dominic Corrigan, John O'Brien, and even Graves in Dublin – took up elements of these themes. If they generally avoided 'rights' language, they did recognise fever as jointly natural and cultural. Nature was ascendant in 1816–17 – weather-related crop failure caused disease – while culture dominated in 1815 and again in 1825–6: there was food, but insufficient employment and buying power.

While only a few addressed policies for fever prevention, many acknowledged that it was the task of doctors, dispensaries and fever hospitals to correct the mess left by market fluctuations and natural variability. Even in good years, doctors noted that many patients needed food and rest, not medicine. All the responses to the great famine – soup kitchens, make-work programmes, rent relief, and distribution of free or discounted food, blankets and clothing, were regular medically administered responses to pre-famine crises.[43] In 1835, Denis Phelan (1785–1871), apothecary and gadfly medical reformer declared that 'the medical charities are in fact the *poor law* of Ireland'.[44]

Phelan wrote amidst the exhaustive examination of social conditions by the Royal Commission on the State of the Irish Poor, chaired by the Protestant Archbishop of Dublin and ex-Oxford Professor of Political Economy, Richard Whateley. Having begun with the standard view that poverty was due to insufficient incentives, the commission discovered the central role of land policy and of disease, particularly 'fever': judicious support of the ill, disabled and hungry was an enabling condition of prosperity. In local hearings, assistant commissioners asked witnesses, representing all walks of life, whether doctors should manage outdoor relief. Not all doctors welcomed such a trans-medical role, but many were playing it already.

The Commission's complicated plan to meet Ireland's poverty by agrarian reforms and socio-medical support systems appalled Parliament. The alternative, a watered-down version of the English new poor law, which it passed in 1837, was a disaster. Where Whateley anticipated the famine, the new law's creator, George Nicholls, planned workhouse accommodation for only 3–4 per cent of the 2.4 million the Commission deemed at risk.[45]

It was the new law, exacerbating many of the tensions within Irish medicine, that forced the retreat of a political medicine, and accounted for the minimal medical response to the great famine. Irish medicine was riven on multiple axes: lower middle-class provincial doctors (many Catholic) against Dublin's consultant elite (mainly Protestant); physicians and surgeons against each other and against uppity apothecaries. Intrigues were incessant. In 1839, the young evangelical surgeon Henry Maunsell, co-founder of the *Dublin Medical Press* and holder of a short-lived chair of Political Medicine at the College of Surgeons, sought to push the squabbling aside and position Irish medicine to play the role envisioned in the Whateley report. In the rural medical charities, the bones of a poor law already existed, as doctors brokered relief of the joint problems of poverty and illness. Packing the poor into workhouses, by contrast, would starve them of fresh air, one of the few health-bringing factors in their lives.[46] Yet Maunsell was outflanked by his own college, and his reform initiative collapsed.

The new workhouses did promise new jobs for medical officers, however, and it only slowly became clear that these might come at the cost of existing medical charity posts. These new posts would be administered by a mix of Whig poor-law bureaucrats and local (and, often, significantly Catholic) boards of guardians. The Poor Law Commission's record in England was one of resistance to recognition of the reciprocal relations of poverty and disease, and contempt for medical autonomy and competence. And yet a good many Irish medicos preferred it to the clientilism of the Dublin elite.[47]

Realistically, then, Irish doctors faced the prospect of either acquiescing to poor-law rule or reframing the scope of public medicine to keep it clearly outside the poor-law orbit. That could be done by sharply distinguishing the medical role – of repairing idiopathic accidents – from any social and economic policy. From 1839 on, Parliament was regularly presented with rival bills to reform Irish medical charities and clarify their relation to the poor law. While the Dublin elite largely succeeded in its redefinition, it lost control of the charities, with the exception of the county infirmaries, in 1851.

The consequence was that in 1845–7, a government looking to tap medical expertise for famine relief would have had difficulty figuring out who represented Irish medicine. And, having lost their bid for significant control of welfare, many leaders of the profession were reluctant even to acknowledge the famine's existence: surely all this famine-and-fever talk was a takeover plot by the poor-law faction?

When the leader of the pro-poor-law party, the prominent Catholic clinician Dominic Corrigan, played the old money or life, famine-fever hysteria card in January 1846, in a pamphlet titled 'Famine and fever as cause and effect', he was much more cautious than he had been when making the argument in 1830.[48] His erstwhile mentor and current opponent, Graves, who in 1822 had urged 'a liberal supply of money' as the 'obvious means' to combat fever-inducing anxiety in Galway, now wrote of Corrigan's famine-fever correlation as dangerous 'mischief'.[49] Attacking Corrigan at length in letter in the august *Dublin Journal of Medical Science*, Graves now explained fever wholly in terms of contagion.[50] It might be coincidental with famine, but the one problem was medical and the other not. Indeed, the fever was caused (and could be blamed) on efforts to fix the social problem: 'the Irish epidemic of 1847 had its origin in the congregating together large masses of people at public works and at depots for the distribution of food, and in the overcrowding the workhouses'.[51] How were starving people to be safely fed? Graves need not say. From his standpoint, medical and social were severed. Did he have second thoughts? He is famous for suggesting as his own epitaph: 'He fed fevers'. Certainly, *within the hospital*, he did.

Thus for contingent reasons Irish medicine self-marginalised in the early1840s. How much further might it have gone? The Whateley recommendations, which would have medicalised welfare, nonetheless distinguished the reactive – medicine – from the preventive – land reform. But could restructuring make Ireland healthy? The hidden succouring had perhaps mitigated the most pathetic aspects of poverty and disability, but done little to prevent it. And yet, even the most expansive of the doctors, Grattan, was uneasy in the role of development economist. Part of the failure of fever hospitals, he had argued, was due to the culture of dependency they generated. The cure was economic development: Grattan's land tax would capitalise cottage industry. And yet the very industry he imagined – hand-loom weaving – was falling victim to Yorkshire's mills. Grattan did acknowledge the most prominent diagnosis of Ireland's ills: overpopulation. But he backed off. If an uncultivated acre remained, Ireland was not overpopulated, and anyway, you could not stop people having sex.

Increasingly, Grattan himself wrote from the front lines. In 1826, he left Dublin for a life as a Kildare squire. He did not give up reform, yet like many doctors, he was politically naïve. His securing of the Edenderry Guardians' endorsement for his tax proposals recalls declarations of university towns as nuclear-free zones: what they express most clearly is alienation from mass politics.[52]

3.6   'Bogs'

In pre-famine Ireland the front line of health-building reform was not cottage industry, however, but bog drainage. Testifying to the 1844 Devon Commission on Irish land use, squire Grattan presents as a progressive and successful landowner, but the first thing that commission wanted to know was 'had he drained his bogs'?[53] (He had, in part.) Proposals to fix Ireland by fixing its environment were plentiful, and almost exactly coincident with the rise of the fever hospitals. Beginning with the bog commission of 1809, engineers like Richard Griffith surveyed and classified waste lands. Griffith, as John Stuart Mill reported, had found 6,290,000 acres of wastelands in Ireland. Of these, by conservative estimate, 3,755,000 were improvable, though only a third of those were potentially arable – the others could serve only as pasture.[54] As in America, new land would augment production, health and citizenship; the vision was of hardy small-holders ditching their way to health, sustainability, and even, perhaps, upward mobility. In island Ireland, such reclamation was also the alternative to a Malthusian self-correction.

Usually, bog drainage was to be accomplished by some quasi-public commission and financed by revolving loans. Seventeenth-century fen drainage was a model of a successful enterprise of this sort; Highland Scotland provided a more immediate example of the infrastructural neutralisation of a potentially rebellious population. Both Whig and Tory governments made such schemes centrepieces of famine-response, but even in emergency they could not pass substantive bills. However entrancing the vision of happy peasants on drained bogs, the bills seemed invariably to benefit some landlords more than others, or worse, to threaten property itself: in some, under-used land would revert to the state.[55]

But finally, some within Ireland recognised that to see the problem as technical and practical was an unwelcome distraction from seeing it as political. A preoccupation with blame trounced the analysis of cause, and left responsibility ambiguous. Indeed, a focus on the control of land would reveal how deeply those who protested the results of an agroeconomic order of many-layered subdivision of property with the ultimate exploitation of the poorest, were themselves implicated in it. Daniel O'Connell was called on the carpet as a middleman by a *Times* reporter and bog-drainage advocate early in the famine.[56] The poor too, who were to benefit, were ambivalent. Mill might tax his great brain working out the minimum landholding to transform cottier into upwardly mobile, market participating, peasant proprietor, but many rightly translated improvement as higher rent and lower demand for labour, especially where subsistence arable was to become market pasture. Often, the rational response was sabotage: to welcome the wages from improvement projects but to ensure that improvement never improved.[57] The upshot then was that the very location of the problem as agroeconomic was itself politicised. Technical change, bringing with it socio-economic improvement and presumably a better standard of health was perceived not as liberation, but as a substitute for political justice. They were not wholly wrong. On the other side of the Irish Sea, it is clear that the claim of better management was a conspicuous part of the justification for Westminster rule. Bog drainage was a great white hope; it would prove that able administration trumped political liberty.

As with the central role of medicine in the management of Ireland, bog reclamation has received little attention. Writing in 1950, K. H. Connell found that there had been a good deal of small-scale reclamation, but bogs – like human numbers, pathogen transmission and levels of nutrition – have remained tangential in an historiography preoccupied with culture and nation.[58]

3.7 Conclusion

Either there was or there wasn't an agroecological praxis by which Ireland's population might have been sustained in adequate nutrition and health. That few have made that *the* central question is disturbing. Some economic historians have claimed at least a part of this territory; they have sometimes struggled to defend their approach against others for whom contingency and possibility are less central than tragedy, for whom motive and sensitivity are less important than effectiveness; which measures heroism as perseverance in the face of hopelessness (rather than as logistical competence); and for whom the perpetuation of memory – including martyrdom – needs no defence. The alternative, in which Ireland is a seamless natural and national entity of happy poverty, rather than the site of pervasive, multi-dimensional agroecological conflict often experienced by many real human beings as a mix of infection and malnourishment, has often been attractive both in England and in Ireland.

I end with a contrast. On one side Thomas Wakley, *Lancet* editor and MP, rising in the House of Commons in early 1846. Wakley, rare as a medical MP, expressed astonishment that no Irish member saw need for immediate famine relief. What was needed to stop fever was food – 'immediately, liberally, without stint, and without restraint'. 'All medical men' knew that, he insisted; why, he wondered, were Irish members not pressing this truth on the authority of Irish doctors?[59]

Wakley's puzzlement has been echoed by political historians: O'Connell's repeal party was indecisive in its early response.[60] Famine, and behind it, welfare, were not its axial issues: representation of Ireland's parlous heath and probable dependence on imported food (and on money to buy it) fit poorly with efforts to portray a nation able to handle its own affairs. A common view in which Ireland's primary identity is political and cultural has been largely calibrated to O'Connell's achievement in creating a party that would support repeal of the act of union, however potentially conflicted it might be on other matters. Ultimately (and ironically), through Mitchel, the famine would be enlisted to support O'Connell's cause, even though he feared its valence would be precisely the opposite. In the process, demographic diagnosis – whether by Wakley, Corrigan, Grattan or Mill – was downplayed: whether defined as biospheric events (blight) or failures to act (insufficient supply of food), the causes that ended lives in Ireland in the late 1840s were to be seen as exogenous rather than inherent, a product of what happened to Ireland rather than what Ireland was.

On the other side, the great table of Ireland's disease history appended to the 1851 census by William Wilde, ophthalmologist to Dublin society, medical editor and antiquary. What an odd census this was – to the pages of tables translating anarchy into order, Wilde appended a comprehensive review of famine and fever from the ancient chronicles of the island. He hibernicised both. If both were there as far back as the records went, their cause was the island itself; there can be no useful talk of blame or responsibility. Neither penal laws nor Act of Union was implicated, nor was there any genocidal conspiracy. Grattan, and later Mitchel, erred in failing to incorporate the chthonic power of *longue durée*.[61]

In presenting these deadly events as natural facts, Wilde overlooked political and agroecological contexts. He was a chronicler, not an analyst. Yet one cannot accuse him of inventing the image of Ireland as a land of fever and famine, either as a demographic or as a cultural phenomenon. However frequently they might have occurred over millennia, hunger and 'fever' were occurring sufficiently regularly in nineteenth-century Ireland to be registered as matters needing cultural assimilation as well as some manner of societal response. A character in William Carleton's 1846 famine novel, *The Black Prophet,* (set in the famine/epidemic of 1816–19, which Carleton calls the 'great famine'), observes: 'there's not upon the face of the earth a counthry [*sic*] where starvation is so much practised, or so well understood. Faith, unfortunately, it's the national divarsion wid us'.[62] To represent famine as a form of regional or ethnic entertainment is remarkable; whether it is more or less outrageous than making the starving into pawns serving a political agenda which gains strength in proportion to the magnitude of mortality is debatable.

Fever too, became less disease than Irish trademark. Wilde was an early leader of the Celtic revival. Largely the work of an Anglo-Irish cadre, this late nineteenth-century myth-making excursus would further the agenda of cultural nationalism, though sometimes at the cost of reinforcing a kitsch Ireland of pigs and paddies – and ghosts. Out of that came the best-known allusion to Ireland's fever, set 'in Dublin's fair city, where girls are so pretty'. We are told of a Mary (or Molly) Malone (or Mallon) who 'died of a fever', and that 'no one could relieve her' (or, sometimes, 'save her'). Over the years, a popular song has become icon; Molly is to Dublin as Marianne is to France. It appears that the song is not traditional, appearing in the 1880s, the work of a Scot.[63] Set in no particular age of Irish history (though her sculptor appears to put in the late eighteenth century), her Molly's plight comes to stand for all of it. Leaving aside whether Dublin's girls were especially pretty, the song recasts a notoriously dirty city as 'fair' and makes poverty, in the hand to mouth life of a street vendor, ideally and emblematically Irish. We are not to care about the precise form of 'fever' (or, better for the rhyme, 'fayhver') that kills Molly. According to their own records, Dublin's united medical community – here collected as 'none' – actually had good success with this 'fever', a nosological mix of exhaustion and acute infections. Of course, had they had actually managed to cure Molly, the song would not work: her ghosthood is crucial since memory is primary.

This happy, boisterous song is rarely occasion for reverie on all real Mollys, who are cast out by the Molly of myth. Their cries, and sometimes, their more physical forms of importuning, were not welcome to the Dublin elite in 1817 and again 30 years later. The fever hospitals were part of the social safety net (or pressure-release valve) that these Dubliners, and those in other parts of the island, had created for those Mollys, who included both native Dubliners and refugees from the countryside, whose agroeconomic order could not always sustain its population. If the statue of Molly were a black jockey on a lawn, most would be appalled. But the myth-history of which she is iconic does celebrate poverty and disease.

Notes

1. The starkest rendering, which frames this chapter, is the Bonnifield (Webb, Malin)-Worster controversy on the dust bowl of the 1930s. See W. Cronon (1992). 'A place for stories: nature, history, and narrative', *Journal of American History*, lxxviii, 1347–76.
2. The Annalistes' approach has been important for environmental historians. D. Worster (1990). 'Transformations of the Earth: toward an agroecological perspective in history', *Journal of American History*, lxxvi, 1087–1106; R. White (1990). 'Environmental history, ecology, and meaning', *Journal of American History*, lxxvi, 1111–16, at 1112.
3. See A. Bourke (1993). 'The visitation of God?', in J. Hill and C. Ó Gráda (eds). *The Potato and the Great Irish Famine* (Dublin: Lilliput Press). Bourke's major sources, like Redcliffe Salaman, *The Potato*, and, behind it, Large's *The Advance of the Fungi*, have been rarely used in recent work. But see J. S. Donnelly Jr. (2001). *The Great Irish Potato Famine* (Stroud: Sutton).
4. C. Ó Gráda (1997). 'The great famine and other famines', in C. Ó Gráda (ed.). *Famine 150: Commemorative Lecture Series* (Dublin: Agriculture and Food Development Authority/UCD), pp. 146–7.
5. A. Sen (1981). *Poverty and Famines: An Essay in Entitlement and Deprivation* (Oxford: Clarendon). Commentators on Sen disagree whether only some famines mark failed entitlement or all do. Sen initially highlighted cases in which famines occurred without any fall in harvest; yet even in cases where there are deficits, issues of failed obligation may still arise.
6. Worster, 'Transformations', p. 1088. See also C. Ó Gráda (1984). 'Malthus and the prefamine economy', in A. Murphy (ed.). *Economists and the Irish Economy from the Eighteenth Century to the Present Day* (Dublin: Irish Academic Press/Hermathena/Trinity College Dublin), pp. 75–95.
7. Worster, 'Transformations', p. 1100.
8. Quoted in C. Woodham-Smith (1962). *The Great Hunger: Ireland 1845–1849* (New York: Harper and Row), pp. 158–9, 203–4. See also W. MacArthur (1956). 'Medical history of the famine', in R. Dudley Edwards and T. Desmond Williams (eds). *The Great Famine: Studies in Irish History 1845–52* (Dublin: Irish Committee of Historical Sciences/Browne and Nolan), pp. 263–315; L. Geary, 'What people died of during the famine', in C. Ó Gráda (ed.). *Famine 150*, pp. 95–117; J. Mokyr and C. Ó Gráda (2002). 'Famine disease and famine mortality: lessons from the Irish experience', in T. Dyson and C. Ó Gráda (eds). *Famine Demography: Perspectives from the Past and Present*, (Oxford: Oxford University Press), pp. 19–43; D. Arnold (1988). *Famine: Social Crisis and Historical Change* (Oxford: Basil Blackwell), pp. 22–3.
9. S. J. Connolly (1989). 'Union government, 1812–23', in W. E. Vaughn (ed.). *A New History of Ireland, 10 Vols. V. 5 Ireland under the Union I, 1800–1870* (Oxford: Oxford University Press), pp. 48–74, at 61–2. Cf. C. Maxwell (1946). *Dublin under the Georges, 1714–1830* (London: Harrap), pp. 120–1; Ó Gráda, 'Malthus and the prefamine economy', p. 83.

10. One school, known as the synergists, holds that both acute starvation and chronic malnutrition weaken resistance. Their critics, citing research that suggests that microbes prefer healthy hosts, must treat links between famine and disease as incidental. Thus: during episodes of hunger, social systems crumble: people migrate and cannot maintain hygienic standards, hence pathogens find new victims. See A. G. Carmichael (1983). 'Infection, hidden hunger, and history', in R. I. Rotberg and T. K. Rabb (eds). *Hunger and History: The Impact of Changing Food Production and Consumption Patterns on Society* (Cambridge: Cambridge University Press), pp. 51–66; R. Dirks (1993). 'Famine and Disease', in K. Kiple (ed.). *The Cambridge World History of Human Disease* (Cambridge: Cambridge University Press), pp. 157–63; A. Hardy (1988). 'Urban famine or urban crisis? Typhus in the Victorian city', *Medical History*, xxxii, 401–25. Few have taken the position asserted by M. Davis (2001). *Late Victorian Holocausts: El Niño Famines and the Making of the Third World* (London: Verso), pp. 21–2, that the distinction between infective and agroecosystemic identity should not be pursued, at least in part because 'famine and epidemic mortality are not epistemologically distinguishable'.

11. C. Hamlin (2008). 'Is all justice environmental?', *Environmental Justice*, I, 145–7.

12. The post-Foucauldians are not alone. Cf. the critique of moral philosophers in A. MacIntyre (1999). *Dependent Rational Animals: Why Human Beings Need the Virtues* (Chicago: Open Court). See also Worster, 'Transformations', p. 1090; Cronon, 'Stories', p. 1349.

13. This has often been the strategy. See T. Meade and D. Serlin (2006). 'Disability and history, editors' introduction', *Radical History Review*, xciv Winter, 1–8.

14. J. Vernon (2005). 'The ethics of hunger and the assembly of society: the techno-politics of the school meal in modern Britain', *American Historical Review* cx, 3, pp. 693–725, at 694. Vernon contrasts his approach with those for whom 'hunger is a self-evident *material* condition that can be affirmed by ahistorically reproducing the forms of nutritional representation by which it was measured'. Vernon groups those like Sen, who focus on entitlement rather than absolute scarcity, as holding a view in which hunger is 'a natural, material condition'.

15. On representations of the hungry, see S. McLean (2004). *The Event and its Terrors: Ireland, Famine, Modernity* (Stanford: Stanford University Press); M. Kelleher (1997). *The Feminization of Famine: Expressions of the Inexpressible* (Durham NC: Duke University Press).

16. M. Ellman (1993). *The Hunger-Artists: Starving, Writing, and Imprisonment*, (Cambridge MA: Harvard University Press). Though focusing on elected starvation, Ellman regards hunger as 'a form of speech' (p. 3). She pushes the nature-culture boundary far towards culture: 'But the fact that hunger endangers our existence, whereas celibacy merely embitters it, does not mean that hunger is more "natural" than sex, less resonant with cultural signification' (p. 36). Both statements may be true; yet they can distract by focusing on hearers rather than speakers, by assuming 'agency', and ignoring pathology.

17. Davis, *Late Victorian Holocausts*, p. 21. Davis notes that the very identification of famines registers chronic rural malnutrition as normal.

18. A. Nicholson (1857). *Ireland's Welcome to the Stranger*, Maureen Murphy (ed.) (Dublin: Lilliput Press, 2002), pp. 9, 17, 20.
19. Such views have been easier to see in their evangelical terms following B. Hilton (1988). *The Age of Atonement: The Influence of Evangelicalism on Social and Economic Thought, 1795–1865* (Oxford: Clarendon); P. Mandler (1990). *Aristocratic Government in the Age of Reform: Whigs and Liberals, 1830–1852* (Oxford: Clarendon Press); P. Gray (1999). *Famine, Land, and Politics: British Government and Irish Society, 1843–1850,* (Dublin: Irish Academic Press).
20. A. Trollope (1994). *Castle Richmond*, Max Hastings (ed.) (London: Trollope Society), pp. 334–6. See Kelleher, *Feminization*, pp. 39–57; and C. Morash (1995). *Writing the Irish Famine* (Oxford: Clarendon Press), esp. pp. 42–51.
21. J. Mitchel (1858). *The Last Conquest of Ireland (Perhaps)* (Glasgow: Washburne), p. 107.
22. Donnelly, *The Great Irish Potato Famine*, pp. 212–3; Morash, *Writing the Famine*, pp. 52–75. See also R. Mahony (1996). 'Historicising the famine: John Mitchel and the prophetic voice of swift', and T. A. Boylan and T. P. Foley (1996). 'A nation perishing of political economy?' in C. Morash and R. Hayes (eds). *'Fearful Realities' New Perspectives on the Famine* (Dublin: Irish Academic Press), pp. 131–7 and 138–50.
23. Mitchel, *The Last Conquest of Ireland (Perhaps)*, pp. 97, 115–17, 125, 133.
24. E.g., E. O'Brien and A. Crookshank, with Sir Gordon Wolstenholme (1984). *A Portrait of Irish Medicine: An Illustrated History of Medicine in Ireland* (Dublin: Ward River Press/Royal College of Surgeons in Ireland); S. Taylor (1989). *Robert Graves: The Golden Years in Irish Medicine,* (London: Royal Society of Medicine).
25. P. Froggatt (1989). 'The response of the medical profession to the great famine', in E. M. Crawford (ed.). *Famine: The Irish Experience 900–1900: Subsistence Crises and Famines in Ireland* (Edinburgh: John Donald), pp. 134–56, at 135–6, 139; R. B. McDowell, 'Ireland on the Eve of the Famine', in Edwards and Williams, *The Great Famine*, pp. 1–86, at 31; T. P. O'Neill (1973). 'Fever and public health in prefamine Ireland', *Journal of the Royal Society of Antiquaries in Ireland*, ciii, pp. 1–34; R. D. Cassell (1997). *Medical Charities, Medical Politics: The Irish Dispensary System and the Poor Law, 1836–1872* (Woodbridge, Suffolk: The Royal Historical Society/The Boydell Press), esp. pp. 4–15; L. Geary (2004). *Medicine and Charity in Ireland, 1718–1851* (Dublin: University College Dublin Press).
26. MacArthur, 'Medical History'; Froggatt, 'Response'; L. Geary (1996). '"The late disastrous epidemic": medical relief and the great famine', in C. Morash and R. Hayes (eds). *'Fearful Realities': New Perspectives on the Famine* (Dublin: Irish Academic Press), pp. 49–59.
27. Anon. (1846). 'The new Irish fever bill', *Dublin Medical Press (DMP)* xv, 188–90; (1846). 'Progress of the famine fever', *DMP*, xvi, 92–3; (1846). 'An Irish fever bill', *DMP*, xv, 173; (1847). 'The fever act amendment bill', *DMP*, xvii, 220; (1847). 'Progress of the fever bill', *DMP*, xvii, 251–2; (1847). 'The board of health', *DMP*, xviii, pp. 204–5; (1847). 'More medical legislation for Ireland', *DMP*, xvii, pp. 204–6; R. Graves (1847). 'A letter to the editor of the Dublin Quarterly Journal of Medical Science, relative to the proceeding of the Central Board of Health of Ireland', *Dublin Quarterly Journal of Medical Science*, iv, pp. 513–44.

28. Excluding poor law infirmaries. Oldest were the 40 county infirmaries (built following an act of 1765). There were also 661 dispensaries (general act 1805), and about 100 fever hospitals (general act 1807; most built after the wake of the 1817–19 epidemic). There were also 30 Dublin hospitals.
29. Documents on the Fever of 1816–1818, National Archives of Ireland (NAI) OP 474/3. On central European medical police precedents see P. Carroll (2002). 'Medical police and the history of public health', *Medical History*, xlvi, 461–94; C. Hamlin (2008). 'The fate of "the fate of medical police"', *Centaurus*, l, 63–9.
30. N. Gash (1961). *Mr. Secretary Peel: The Life of Sir Robert Peel to 1830* (London: Longmans), pp. 219–25.
31. Documents on the Fever of 1816–1818, NAI OP 474/30, 35, 45–49.
32. Documents on the Fever of 1816–1818, NAI OP 474. 54, 57, 60, 63; cf. 3, 56, 59.
33. Documents on the Fever of 1816–1818, NAI OP 474/61–62.
34. Such phenomena are consistent with Brill-Zinsser disease, a mild reemergence of a latent typhus acquired in infancy, hospitalisable 'fever' embraced a far wider range of illnesses, including, very likely, scurvy and other manifestations of malnutrition. M. Crawford (1988). 'Scurvy in Ireland during the great famine', *Social History of Medicine*, i, 281–300.
35. J. O'Brien (1820). 'Medical report of the fever hospital and house of industry, Cork Street Dublin, for 1820', *Transactions of the Association of Fellows and Licentiates of the King and Queen's College of Physicians in Ireland [TCPI]*, iii, pp. 448–505, at 477.
36. L. Kennedy, P. S. Ell, E. M. Crawford and L. A. Clarkson (1999). *Mapping the Great Irish Famine: A Survey of the Famine Decade* (Dublin: Four Courts Press), p. 118; MacArthur, 'Medical', pp. 279–80.
37. J. Scott (1985). *Weapons of the Weak: Everyday Forms of Peasant Resistance* (New Haven: Yale University Press).
38. H. Maunsell (1839). 'Sketch of a discourse on political medicine', *DMP*, i, pp. 65–6.
39. R. Grattan (1817). 'Medical report of the fever hospital in Cork Street, Dublin, for the Year 1815', *TCPI*, i, pp. 435–94, at 458.
40. R. Grattan (1820). 'Medical report of the fever hospital in Cork Street, Dublin, for the Year ending 4th January 1819', *TCPI*, iii, pp. 316–447, at 324–9.
41. Grattan, 'Medical Report ... for ... 1819', p. 373.
42. R. Grattan (1826). *Observations on the Causes and Prevention of Fever and Pauperism in Ireland* (Dublin: Hodges and McArthur), pp. 3–4.
43. Fullest are the medical reports of the Royal Commission on the State of the Poor in Ireland, *First Report, Appendix B, P.P.* 1835, v. 12.
44. D. Phelan (1835). *Statistical Inquiry into the Present State of the Medical Charities of Ireland* (Dublin: Hodges and Smith), p. 9.
45. P. Gray (2009). *The Making of the Irish Poor Law, 1815–43* (Manchester: Manchester University Press), p. 165.
46. H. Maunsell (1838). *The Only Safe Poor Law Experiment for Ireland, A Letter to the Rt. Hon. Lord Viscount Morpeth* (Dublin: Fannin).
47. Geary, *Medical Charities*; C. Hamlin (1998). *Public Health and Social Justice in the Age of Chadwick* (Cambridge: Cambridge University Press).

48. Corrigan (1830). 'On the epidemic fever in Ireland', *Lancet*, ii, pp. 569–75, 600–3; Corrigan (1846). *On Famine and Fever as Cause and Effect in Ireland; with Observations on Hospital Location and the Dispensation of Outdoor Relief of Food and Medicine* (Dublin: J. Fannin).
49. R. Graves (1824). 'Report of the fever lately prevalent in Galway and the West of Ireland', *TCPI*, IV, 408–38, at 423; R. Graves (1884). *Clinical Lectures on the Practice of Medicine*, 2nd edn Dr Neligan (ed.) (London: New Sydenham Society), i, p. 106.
50. Graves, 'Letter to the editor'.
51. Graves, *Clinical Lectures*, i, p. 107.
52. R. Grattan (1845). *How to Improve the Condition of the Industrious Classes* (Dublin).
53. Royal Commission on the State of the Law and Practice with Respect to the Occupation of Land in Ireland [Devon Commission]. P.P.; 1845. ev, pt III, v. 21 [657.], 652–9; K. Kiely (2006). 'Poverty and Famine in County Kildare, 1820–1850', in W. Nolan and T. McGrath (eds). *Kildare: History and Society* (Dublin: Geography Publications), pp. 493–534; W. Nolan (2006). 'The land of Kildare: Valuation, ownership and occupation, 1850–1906', in ibid., pp. 549–84, at 562–66.
54. J. S. Mill (1986). 'The condition of Ireland', in *Newspaper Writings, January 1835–June 1847*, A. and J. Robson (eds). *The Collected Works of John Stuart Mill*, 39 vols, v. 24 (Toronto: University of Toronto Press/Routledge and Kegan Paul), pp. 962–4.
55. Gray, *Famine, Land, and Politics*, 81, 153; T. C. Foster (1847). *Letters on the Condition of the People of Ireland*, 2nd edn (London: Chapman and Hall), pp. 229–53; L. Kennedy (1829). *On the Cultivation of the Waste Lands in the United Kingdom* (London: Ridgway); W. B. MacCabe and F. Jeffrey (1847). 'Measures for Ireland', *Dublin Review*, xxii, pp. 230–60.
56. Foster, *Letters on the Condition of the People of Ireland*.
57. R. B. McDowell, 'Ireland on the eve of the famine', in Edwards and Williams (eds). *The Great Famine*, pp. 1–86, at 7.
58. K. H. Connell (1950). 'The colonization of waste land in Ireland, 1780–1845', *Economic History Review*, n.s. (iii), pp. 43–71.
59. 'Fever and famine (Ireland)' in *Hansard's Parliamentary Debates, 3rd Series*, 84, 9 March (1846), pp. 981–2.
60. K. B. Nowlan, 'The Political Background', in Edwards and Williams (eds). *The Great Famine*, pp. 131–206.
61. E. M. Crawford (1989). 'William Wilde's table of Irish famines, 900–1850', in Crawford (ed.). *Famine: The Irish Experience 900–1900* (Edinburgh: John Donald), pp. 1–30.
62. W. Carleton (1847). *The Black Prophet: A Tale of the Irish Famine* (edn Poole: Woodstock Books, 1996), p. 348.
63. http://www.contemplator.com/ireland/cockles.html, accessed 22 October 2009.

4

The *Handbuch der Hygiene*: A Manual of Proto-Environmental Science in Germany of 1900?

Dieter Schott

4.1 The *Handbuch der Hygiene* – aims and purposes

In 1893 the chemist and physician Theodor Weyl, working at the Technical University of Berlin-Charlottenburg, started a major publication venture. Together with the renowned publisher Gustav Fischer in Jena he announced the launch of a massive handbook project on 'Hygiene'.[1] Remarkably, in his preface to the first volume, Weyl above all emphasised the impartiality of his project: 'The *Handbuch der Hygiene* does not put itself in the service of a particular school or some, even if particularly eminent personalities'.[2] He wanted to include all strands of science and welcomed representatives of all different schools as collaborators to the handbook. This impartiality had been achieved, Weyl argued, because the authors were largely recruited from the ranks of architects and engineers not partisan to particular hygienic dogmas.

What now was the purpose of the *Handbuch*? Weyl claimed to have identified a deficit in academic training in the field, which intersected between the activities of physicians and engineers. He noted a lack of sanitary engineers, at least compared to Britain and the US, and judged the academic training in hygiene received by physicians in Germany and Austria as not really qualifying the physicians employed by the state or by local government to cope with the enormous range of different tasks with which they were confronted in their daily work. Weyl linked this critique of insufficient academic training with a general critique of health policy and of the administrative structure of health administration. He demanded that health issues in Prussia should be separated from the Ministry of Culture and should be part of the portfolio of the Ministry of the Interior or – even better – be taken care of by a separate Minister of Health who – according to Weyl – should naturally be a physician.[3]

The *Handbuch*, Weyl underscored, should thus be directed at the intersection between medicine, engineering and civil administration, and should offer knowledge and practical advice to all those practitioners.

4.2 What was 'hygiene' around 1900? The contents of the *Handbuch*

What exactly did 'Hygiene' mean, and what was its significance to the late nineteenth-century German public? Since Max Pettenkofer had been appointed to the first chair of hygiene at Munich University in 1865, hygiene, much in line with medicine as a whole, had undergone a comprehensive process of scientification.[4] From its mid-nineteenth-century Chadwickian beginnings, public health was rooted in social policy far more than in medicine.[5] Pettenkofer and his rapidly growing number of disciples and followers had developed hygiene into an experimental science, minutely charting and analysing the incidence of epidemic as well as endemic diseases in particular sites and locations and linking this to a wide range of environmental factors such as soil composition, humidity, ground-water level etc.[6] By elaborating this multi-factorial theory of causation of epidemic diseases, particularly those such as cholera, he created a scientific basis for the huge investments in sanitary engineering undertaken by German cities from the late 1860s.[7] And by developing a wide range of experiments he expanded the knowledge base of hygiene, for instance in terms of respiratory needs, specifying the amount of air an adult needed. Thus he provided the basis for engineers to plan not only sewage systems but also housing and the layout of towns according to scientifically established biological and hygienic requirements.[8]

This new knowledge, claiming to be based on strict scientific methods, was popularised and reached a wider public through societies such as the 'Deutsche Verein für öffentliche Gesundheitspflege' (German Society for Public Health), founded in 1873. Here physicians collaborated with engineers, statisticians and architects in debates over public demands for certain measures to be adopted by the state and particularly by German cities. This society was probably the most influential public health lobby in the 1870s and 1880s.[9] Its recommendations and resolutions carried considerable weight, not only with the academic sphere but also with civil servants charged with coping with the difficult problems of urban growth and high urban mortality rates.

The rise of bacteriology as a completely different approach to disease causation now presented itself from the 1880s as a major challenge to the hegemony of hygiene as taught by Pettenkofer and his followers.

Pettenkofer's hygiene took a comprehensive range of environmental factors into account, and its strategies focussed on preventive action which would avoid or diminish the emergence of conditions likely to foster the production of noxious miasma, such as comprehensive groundwater control and the evacuation of waste water. Bacteriology, in contrast, focused on the identification of a single causative germ and the search for directly targeted counter measures, such as disinfection or creating immunity by vaccination, a practice, which had already been successfully established for a few diseases such as smallpox.[10]

The Hamburg cholera epidemic of 1892, which caused almost 8000 fatalities, is commonly held to have dealt the death blow to classic environmental hygiene in the Pettenkofer tradition.[11] It fully exposed the catastrophic shortcomings of sanitary arrangements in Hamburg legitimated by these concepts – i.e., the rejection of the possibility of cholera being transmitted by drinking water. However, in terms of the scientific debate among medical doctors and practitioners of public health, we can observe that the result of the Hamburg cholera epidemic was not a full and complete victory of germ theory but rather a 'negotiated closure'.[12] And this ambivalent outcome, the notion that hygiene in the Pettenkofer sense stills holds its place is clearly reflected in the content of the early volumes.

Volume 1, which was published in 1893 shortly after the cholera in Hamburg, dealt with what one could term 'basics' of hygiene: the history of public health in 'civilised' nations,[13] the externally given factors such as soil and climate, and clothing as the man-made immediate environment of the body (see table 4.1). In the essay on 'soil' the author Josef von Fodor presented a ringing dedication to Pettenkofer, despite what just had recently happened in Hamburg. Somewhat unexpectedly we also find an essay on 'Climate and hygiene in the tropics' which was due to the then intense concern, in Germany and elsewhere, to acquire colonies in tropical latitudes and to investigate the possibilities and problems of white European settlement particularly in Africa, which had been so impenetrable in earlier times due to the disease barrier.[14] If 'climate' and 'soil' as basics of hygiene seem to suggest geo-determinist modes of thinking, then tropical hygiene obviously strove for possibilities to overcome the obstacles of tropical climate by a variety of technical and cultural means. The second part of volume 1, dealt with drinking water and its provision and control. Clean drinking water, quite in line with traditions since Edwin Chadwick, was considered the essential medium of promoting or endangering hygiene and thus was attributed a place of particular importance in the first volume.

Table 4.1 Table of contents of *Handbuch der Hygiene* volumes 1–10

Vol. no.	German title	Translation of title	1st half volume	2nd half volume
1	(no specific title)		History, soil, climate, clothing, 'climate and hygiene in the tropics'	Drinking water: provision and control
2	Städtereinigung	Urban Sanitation	Sewage systems, irrigation farms, agricultural use of faeces, river pollution, street hygiene, waste disposal	Burial systems, removal of animal corpses
3	Einzelernährung und Massenernährung	Individual and Mass Nutrition	Human metabolism/nutrition/ meat and meat inspection/human diet/mass nutrition	Properties of individual foodstuffs and drink/cans and packages of food and drink
4	Bau- und Wohnungs-hygiene	Construction and Housing Hygiene	Impact of housing on health, housing statistics, lighting in general, gas lighting	Circulation of air, heating, housing inspection, hygiene of town planning (Stuebben), the house for living, legal provisions on low-price housing, bacteriology and biology of housing
5	Bau und Betrieb der Krankenhäuser/ Hygiene der Gefängnisse	Construction and Operation of Hospitals/Hygiene of Prisons	Hygienic construction of hospitals; administration and operation of hospitals	Hygiene of prison system

6	Spezielle Bauhygiene	Special Hygiene of Construction	Public baths, asylums, theatres, ships and railway hygiene
7	Schulhygiene und öffentlicher Kinderschutz	School hygiene and public protection of children	Handbook of school hygiene: construction, organisation of school routines, Public protection of children: demographic patterns, birth and perinatal care, nutrition of children, disease prevention
8	Gewerbehygiene	Industrial hygiene	General industrial hygiene and factory legislation, hygienic care for female workers, mechanical devices to prevent accidents, ventilation of workshops. Hygiene of millers, bakers and confectionaries; hygiene of tobacco workers, hygiene of chemical industry, hygiene of glass workers
9	Aetiologie und Prophylaxe der Infektionskrankheiten	Aetiology and prevention of infectious diseases	Immunity, parasitology Epidemiology, public measures against infectious diseases
10	Hygiene der Prostitution und venerischen Krankheiten	Hygiene of prostitution and venereal diseases	Hygiene of prostitution and venereal diseases General register to all 10 vols

The second volume followed the well-trodden paths of public health, dealing with sanitation of cities, sewage systems, irrigation farms, the agricultural usage of faeces, river pollution, street hygiene and waste disposal. The second part discussed burial systems and removal systems for animal corpses; it thus treated human and animal 'waste' as a problem of hygiene. Contrasting with, and complementing volume 2, volume 3 focused on the intake side of human metabolism, on food and drink, including its various abuses or food-related hazards, such as food poisoning through additives, conservation techniques and so on (see table 4.1).

Volumes 4–6 inquired into hygienic aspects of the physical environment beyond individual metabolism. Volume 4 focused on health aspects of the physical environment of the house, on construction issues of a variety of houses, including also issues and instruments of housing reform. The volume also included a comprehensive discussion of town planning in relation to hygiene, presented by the eminent architect and pioneering town planner Josef Stübben.[15] Stübben had just become famous for planning the monumental city extension of Cologne after its fortification walls had been removed in the early 1880s and in 1890 he had published a comprehensive manual *Der Städtebau*, offering a fully fledged handbook on the practice and rules of town planning as they had evolved in Germany by then.[16] The fact that Weyl could attract Stübben as contributor clearly shows that on one hand he was well aware of who were the leading and active authorities in particular fields, and on the other that for somebody like Stübben the *Handbuch* project was attractive enough to participate in (see table 4.1).

Volume 5 shed light on specific issues of catering for the health of people in institutions such as hospitals, prisons and military buildings, where inmates tend to spend a longer time in rather crowded spaces. Volume 6 analysed the 'special hygiene of construction', which reflected ways of building public amenities such as market halls and cattle courts, public baths, asylums, theatres, ships and railways in ways to correspond with requirements of modern hygiene. Volume 7 occupied itself with a topic which would hardly be subsumed under 'hygiene' today: children's protection and school hygiene.[17] Volume 8 inspected work conditions for specific trades particularly exposed to health hazards such as miners, millers, bakers, tobacco workers etc. Volume 9 then focused on issues of medicine in a more narrow sense, discussing the aetiology of infectious diseases and the potential to prevent infection and spread of such diseases. Here the still fairly new science of bacteriology came into its own. Volume 10 was devoted to hygiene of prostitution and venereal diseases. It furthermore comprised an epilogue, noting the death of several contributors and thanking the authors and the publisher, and a compendious general register to all ten volumes. This tenth volume was released in 1901 (see table 4.1).

In the following years the *Handbuch* was augmented by several supplementary volumes (see table 4.2). The articles partly updated earlier chapters, and partly gave additional information on topics hitherto not covered, such as the protection of cities against fire hazard in volume 2. In these supplementary volumes aspects of what we would call environmental pollution figured more and more prominently, as in volume 3's articles on the noise problem in cities, on atmospheric pollution and on the smoke nuisance.

Supplementary volume 4, a tome of over 1000 pages published in 1904, indicated the direction that 'hygiene' was beginning to take in the early twentieth century (see table 4.2). In his preface the editor Weyl, himself author of several substantial articles in this volume, defined 'Social Hygiene':

> as the kind of science, which commits itself to enable each age group to achieve the lowest possible mortality which would be attainable according to natural circumstances.[18]

Weyl's understanding of 'social hygiene' was therefore fairly close to what nowadays would be called social policy; it aimed to target specific social groups in their health condition whereas classical 'conditional hygiene' as developed by Pettenkofer and his disciples in the second half of the nineteenth century addressed itself basically to a general undifferentiated public.[19] Thus social hygiene had to comprise, according to Weyl, issues such as land reform, welfare reform, food policy or housing policy, as they all impacted directly or indirectly on the ability of specific social groups to reach an average age of death comparable to other social groups.

The background to this change of paradigm towards social hygiene can be found particularly in the new focus on tuberculosis as a disease which mass testing in the late nineteenth century had shown to be endemic in large parts of the urban population.[20] Early hopes of finding a vaccine for tuberculosis, a project also pursued by Robert Koch, had vanished. The strategy of totally isolating the infected from the healthy proved unfeasible for financial reasons in the light of the massive rate of (still latent) infection. The new approach was now to make the masses of the population disease-conscious, so that the spread of the disease could be contained without total isolation of all those suffering

Table 4.2 Table of contents of *Handbuch der Hygiene* supplementary volumes

Vol.	German title	Translation of title	1st half volume	2nd half volume
1	(no specific title)		Hygiene of teaching/hygiene of working in compressed air/hygiene of alcoholism	
2	(no specific title)		Protection of cities against fire	Separate sewer system/infection in hair-dresser salons/dental hygiene
3	(no specific title)		Noise in cities/air pollution by industrial companies	Smoke question
4	Soziale Hygiene	Social Hygiene	Urban sanitation/prevention of dangerous diseases/alcoholism/nutrition/ poor welfare/housing/baby care/school children care/welfare for adolescents/ workers' protection/welfare for workers and families/demography and war/ degeneration/history of social hygiene	

from tuberculosis. This approach re-directed the focus of hygiene from the public to the private: In addition to removing filth from streets and public spaces and thus eliminating sources of 'miasma', it now became important to educate housewives in how to clean their homes and how to prevent infected family members from spreading germs.[21] And since tuberculosis was a disease particularly afflicting the poor, raising standards of cleanliness in the homes of the poor became a particular focus of middle-class female charity work. By 1905 1000 private agencies to combat tuberculosis had been established in the German Empire and their focus had shifted from funding regional tuberculosis sanatoria to taking care of patients in their homes and assisting their families. Many middle-class women were engaged in these tuberculosis agencies and it constituted an empire-wide health movement, presided over by the Empress as patroness and the Chancellor, who was honorary president of the Deutsches Zentralkomitee zur Bekämpfung der Tuberkulose ('German Central Committee for the Battle against Tuberculosis'), founded in 1895.[22] This health movement aligned health policies pursued by the state, as in the sickness and invalidity insurance system inaugurated by Bismarck, with middle-class voluntarism by social reformers and the non-socialist women's movement.[23] Concern about tuberculosis was linked with the more general fear that the health stock of the general population would deteriorate due to such diseases, unhealthy living conditions and the higher birth rate of people with a weaker constitution.[24] And in the context of social Darwinist theories about imperialist competition between nations depending on their physical powers, such a prospectus of general health degeneration became a national political issue.

4.3   The 1890s as a time of change and challenges

What now was the wider historical context of the *Handbuch*? Around 1890 the German Empire moved into a different phase of its history.[25] This was most marked in the political sphere. After Wilhelm II had come to the throne in 1888 he finally dismissed the Iron Chancellor Bismarck in 1890; the style of statesmanship and diplomacy practised by Bismarck was not to the ambitious young Emperor's liking. Aiming for colonial aggrandisement and a prominent role in world politics, Wilhelm II did away with the skilful, but tangled diplomatic manoeuvres of Bismarck and set upon a strategy of developing Germany's naval power, thus eventually bringing Germany as an imperial contender up against Britain.

In the domestic sphere the year 1890 also marked major shifts, such as the abolition of the socialist laws, which had made socialist agitation and party campaigning illegal. The newly legal social democratic party (SPD) staged a spectacular comeback at the polls, almost steadily increasing its share of votes after 1890 in a climate of sharp class antagonism.[26] In 1890 the SPD won almost one fifth of the votes at elections to the Reichstag, the Imperial Parliament, and by 1912 this share had increased to over one third.[27] Despite Wilhelm II's ambition to be seen as 'the people's emperor', and despite generous measures of social policy early in his reign the majority of workers kept a sceptical distance from the monarchy and maintained their allegiance to the socialist movement.

For a short period in the early 1890s, under Chancellor Leo von Caprivi, the power base of the Imperial government changed from the Bismarckian coalition of Junkers and coal barons to a more open-minded grouping of modernisers from the new and export-oriented sections of German business, such as the rapidly rising electrical and chemical industry. Far more than Britain, but similarly to the US, Germany's economy was transformed by the second industrial revolution, which shifted growth dynamics and employment to other sectors and regions. From the mid-1890s Germany's economy experienced an almost unbroken period of strong growth which definitely shifted the scales towards industry in the debate over whether Germany was predominantly an agrarian or industrial society. Whereas industrialisation before 1880 had been concentrated where iron and coal predominated, as in the Ruhr region, or where engineering and textiles were prominent as in Berlin or Saxony, now the new industrial sectors in electrical engineering, chemicals and car making emerged in regions which had hitherto hardly been industrialised such as Southern Germany.[28] This new coalition of modernisers also gave an impetus to those middle-class academics and professionals who were traditionally loyal to the Emperor and the Empire but had resented the conservative course and style of domestic politics under Bismarck. For them a 'new course' seemed to open up after 1890; they were deeply convinced that they had a substantial contribution to make towards social peace and the general betterment of society by means of scientific as well as cultural improvements. 'Progress' through science and technology, not so much through politics, would be the common denominator of this group.[29] Although 'progressivism' has hardly been used as an analytical concept for German society, as it was for the US and to a certain extent also for Britain, in Germany there also existed a sizeable community of scholars, scientists, technical experts and academically trained civil servants who were deeply imbued with what I would call 'the progressive gospel'.[30] And in view of the reluctance, particularly of the Prussian elites, to engage with progressive reforms, arenas such as the cities where liberal majorities in municipal assemblies could be persuaded into long-term municipal intervention by progressive leaders, acquired particular importance, not least for improvements in the field of sanitary measures.[31]

In cultural terms, however, the same period was characterised by a rising wave of anti-urbanist and anti-industrialist critique.[32] The rapid urbanisation since the late 1860s had produced urban environments, which were seen by many as inhuman and debasing.[33] The spatial expansion of cities and the commercialisation of the countryside seemed to destroy natural as well as cultural values rooted in traditional society. Among the urban middle classes such critique of modern capitalist and industrial society, often put forward by conservatives such as Wilhelm Heinrich Riehl or August Langbehn, the populariser of Friedrich Nietzsche, did resonate.[34] In contrast to the anti-capitalist critique advanced by Marxist social democracy, this critique did not focus on ownership of 'means of production' or class relations but on lifestyles. Critics denounced the alienation from nature and the blatantly materialist spirit of contemporary society, and demanded far-reaching reforms in the way people lived, arranged their houses, dressed, ate, exercised and spent their leisure time.[35] By the late nineteenth century a wide range of reform movements had emerged, summarised under the loose label of 'Lebensreform' ('life reform').

These reform movements like vegetarianism, anti-alcoholism, dress reform, nudism and many others fundamentally challenged the ways that people lived in German towns and cities and motivated them to search out and develop alternative, healthier and more natural ways and arrangements. Only very small groups of dedicated activists did actually abandon the city and try to establish alternative communities in the country.[36] The large majority remained in cities but gradually adopted and incorporated into their lifestyles some of the postulates of life reform. Under these auspices demands of popular hygiene were paramount: if modern civilisation was counter to nature and led to the moral and physical degeneration of those forced to live in cities, then it was important to leave the city behind as often as possible, for instance on hiking tours or over extended summer vacations. 'Health' acquired an almost religious importance as a secularised value; to lead as healthy a life as possible became a self-legitimating social goal. The attraction of a concept like the garden-city, not so much in the strict version of Ebenezer Howard's social utopia but rather as a leafy, airy and well-planned suburb on the green edges of town, testifies to the widespread desire for more healthy and natural living environments.[37] And this desire gradually diffused from the middle classes to the better-off sections of the working class in the early twentieth century.

The stage on which this 'progressive gospel', shared by professionals, physicians and engineers, was now enacted was the city. It was here where cholera epidemics, paired with a generally high level of mortality and appalling living conditions of the lower classes, had produced a sense of acute crisis. It was here that the aspiring and ambitious younger members of the middle classes, well trained in academic and technical subjects, sought their professional role and their personal advancement. Leading positions in the military and the state bureaucracy, at least in Prussia, were traditionally reserved for the younger sons of gentry. But cities with their considerable autonomy and their growing economic clout were increasingly perceived as offering attractive career opportunities for ambitious young professionals.[38] In the 1870s and 1880s urban administration in many large German cities changed substantially in character and scope. Where before they had been run by honorary members of the local elites with a basically liberal world view, aiming for a low-intervention, low-cost municipality, they now tended to become ever larger bureaucratic and professionalised institutions. And the new mayors, originating from professional backgrounds frequently with law degrees, and often not part of local elites, now were ready and eager to intervene in the social, economic and cultural life of the city in ever more comprehensive ways.[39] In the wake of this professionalisation, cities also adopted entrepreneurial stances, taking water and gas works, power stations and tramway companies into municipal hands.[40]

This 'municipal socialism' also had a hygienic dimension. By gaining a more comprehensive control of the physical environment of a city, the administration could improve and enhance the sanitary condition of its residents in a much more pervasive fashion. By the late nineteenth century, cities had become major players in the field of hygiene: Besides the 'Stadtphysicus', a customary institution in larger cities, municipal administration came to employ a rapidly expanding workforce whose activities impacted on the health of city residents: workers in the slaughterhouses or the municipal dairy, chemists at the municipal laboratory or food control station, staff at municipal baths or market places, engineers operating heating and ventilations systems in large public buildings and so on.[41]

It was really to this wide range of professional and technical specialists in the service of municipalities and counties, engaged in the preservation and improvement of urban sanitary conditions in a very wide sense, that the *Handbuch der Hygiene* was meant to cater, as Weyl had outlined in his editorial preface.[42]

4.4 'Makers' and 'users' of the Handbuch

Let us now take a look who actually contributed to the handbook, as this was clearly a collective endeavour.[43] For 60 of the 80 authors we can identify their professional occupation with reasonable certainty, and show that the large majority came from medicine in a wide sense (see table 4.3). Besides physicians and medical doctors, comprising well over half of authors, the second-largest group came from engineering disciplines, and among those civil engineers, the profession most intimately linked with urban sanitation, dominated. A small but sizable contingent were chemists, as was Weyl himself who additionally had a medical qualification; other small groups came from teaching, agriculture and economics.

In terms of residence of contributors we can identify a clear bias towards authors from Prussia, especially from Berlin (see table 4.4). Including Charlottenburg, then an independent city just outside Berlin with a

Table 4.3 Professional background of *Handbuch* contributors

Profession	Number	% of total
Physicians	32	53
Engineers	19	32
Economists	2	3
Chemists	3	5
Teachers	2	3
Farmers	2	3
Total	60	100

Table 4.4 Residence of *Handbuch* contributors

	Number	%
In the German Empire *of which:*	66	88
Berlin	*29*	*39*
Dresden	*3*	*4*
Hamburg	*4*	*5*
Austria-Hungary	8	11
Others abroad	1	1
All authors with known residence	75	100

high-ranking technical university, two fifths of authors resided in Greater Berlin. Clearly, Berlin featured a range of leading academic and scientific institutions but nevertheless it appears that Weyl preferred local contributors because he knew the scene better there and reckoned that negotiations over necessary revisions of the articles might be facilitated by physical proximity. Only two other cities apart from Berlin boast more than two contributors: Dresden and Hamburg. A sizeable contribution also came from authors resident in Austria-Hungary, while one contributor happened to be living in Paris. Conspicuous by their complete absence are Swiss authors, although the Technical University at Zurich was by then a highly reputed academic institution. Thus the spread of authors was not really representative for German-language scholarship in hygiene around 1900.

In terms of age groups to which authors belonged it is more difficult to come to valid judgments, since we could only find reliable information for 16 of them. Half of those were born in the 1850s, meaning that by the time of planning and realisation of the *Handbuch* they would have been in their forties, well-established in their fields but not necessarily 'grand old men'. In historical scholarship on the German Empire this age cohort has been termed the 'Wilhelmine generation', as Wilhelm II (b. 1859) somehow exemplifies the mentality of this generation, which thought of itself as epigones, inheriting the already unified German national state and striving to enhance its greatness.[44] The remainder were evenly divided between those in their fifties and in their thirties at the time of writing of the handbook. But since these data only cover about one quarter of the total of contributors, they should not be over interpreted. Finally, in gender terms the *Handbuch* was, perhaps unsurprisingly, almost completely a male affair. Only two of the 80 authors were women, including one of the rare women doctors trained in the late nineteenth century.[45]

Weyl, the general editor, was himself a representative of the 'Wilhelmine generation'. Born 1851 in Berlin, he had studied with famous physicians such as Emil Du Bois-Reymond, Felix Hoppe-Seyler and Robert Koch at the universities of Berlin and Strasbourg, earned his PhD in 1877 with a thesis on vegetal and animal proteins, and gained his *habilitation* for physiology in 1879 at the University of Erlangen.[46] He then was employed as specialist in hygiene in England, Hungary and Turkey before obtaining a position as a teacher for hygiene at the Berlin University in 1896. His research dealt with bacteriological, hygienic and physiological issues. Besides editing the *Handbuch* he also published books on aniline dyes and on street hygiene. He also edited a textbook on workers' diseases and another on methods of organic chemistry, later to be continued by Josef Houben.

Weyl clearly knew the scene, and was intimately familiar with what was going on in the various fields constituting the wide orbit of hygiene. On the other hand he did not seem so high profile and so polarising as the exponents of the different schools like Pettenkofer and Koch. This might have been to his advantage for such a venture, and could have been what enabled him to recruit such a vast number and varied range of authors to this project, some quite prominent. And since the practical purpose was to be in the forefront, the *Handbuch* concentrated on giving practitioners a short but still comprehensive overview of the various problems involved in particular fields, including voluminous lists of further reading as well as illustrations, mostly drawn, and graphics.

What was the readership of the *Handbuch*? This is difficult to tell with precision; we do not have figures on circulation and sales. When it was completed in its original ten volumes it represented on several thousand tightly printed pages the scholarship and practical expertise on an enormous range of issues related to health and hygiene. The complete set was sold at a price of 150 Marks in paperback and 175 Marks in hardcover, which is equivalent to 50 daily wages of a worker for the paperback version.[47] While we should not expect to find the handbook on many bookshelves of workers, the price was not so high as to make the purchase an absolute luxury. Furthermore, parts of the handbook, constituting half-volumes, could be bought separately, at prices between 1.80 Marks and 9 Marks. Thus the knowledge disseminated by the authors of the handbook could reach a much wider audience beyond those institutions and libraries that subscribed to the complete handbook. This is supported by the fact that in many public libraries in Germany the handbook is not available as a complete series but in individual volumes and tomes,[48] which at the time of their publication obviously were of particular interest to the library and institution.

4.5 A manual of proto-environmental science?

How far can the *Handbuch der Hygiene* be called a manual of proto-environmental science as stipulated in the title of this paper? I have not discussed 'environment' extensively so far and would like to come back to that issue for this final section.

A first parallel can be identified in terms of the comprehensiveness and universality of the concept of 'hygiene' around 1900. As the table of contents of the *Handbuch*'s volumes shows, hygiene in the sense used at the turn of the century included many issues we would today not normally consider as a proper part of hygiene or health policy. A similar universality can also be traced in debates of the 1970s–90s where 'Umwelt'/environment acquired an equally comprehensive significance in terms of problematising the observed or assumed deterioration of the physical environment. In contemporary society around 1900 fears about impending racial degeneration and the decline of physical capacities of large parts of the population due to adversities in the physical environment such as smoke pollution were omnipresent; 'hygiene' as a guiding vision thus developed a very strong social appeal. Discussing issues in terms of 'hygiene' responded to the mode in which contemporary society constructed its problems and was highly charged with aspirations of social improvement and progress. The hygiene movement, as it is frequently termed in German historiography, had a clear mission, and this mission was spelled out in a particularly lucid way in the *Handbuch* section on school hygiene: the aim was to create among a whole generation the need and the expectation for proper hygienic circumstances for their lifetime, by accustoming them to proper hygienic circumstances at school and by means of the influence school could exert. The essential impetus was, thus, to initiate a process of progressive 'hygienisation' which would endow school children as future citizens and – in a wider sense the whole society – with a deeper sense of understanding of the necessities of hygienic postulates than their parents.[49] Hygienisation thus was understood as a specific manifestation of the civilising process.

Apart from this progressive sense of hygiene as initiating an educational process, hygiene could and did act – and this is my second point – as a nexus and common point of reference for an increasingly wide range of academic disciplines, from medicine over engineering to economics and the emerging fields of sociology and psychology. In a situation where the increasing specialisation of disciplines seemed to segment and disintegrate the totality of lived experiences, the notion of 'hygiene' or the promotion of healthy living provided a focus on which the practical results of these fields could converge. In this integrating function 'Hygiene' has had a similar role in environmental history to that which the concepts of 'environment' and 'ecology' have played since the late 1960s.

The third point is that there was a specific dialectic at work driving 'hygiene' and 'hygiene politics' beyond its initial confines. Given its original motivation, 'hygiene' is clearly anthropocentric, geared towards improving the health of humans, particularly at first in those areas where health seemed most threatened, in industrial cities. The very success of cleaning up these cities, by means of sewage systems and garbage disposal, meant externalisation of waste matter and spelled pollution for the countryside and the rivers. Protagonists like Reinhard Baumeister and Pettenkofer were adamant that cities had a natural right to use rivers as sinks and that rivers had a natural capacity for self-cleansing which would enable them to cope with the sewage.[50] Against an initially fierce, but slowly receding resistance from agriculture, fishery and the Prussian state, cities won out. But by 1900 the disastrous effects of such a strategy, emitting untreated sewage into rivers, became increasingly apparent. Since rivers were not just sinks but also resources, conflicts between cities developed. In a famous dispute between cities using the Rhine, Mannheim as a sink for its sewage and the downstream city of Speyer for drawing drinking water, state authorities increasingly came to oblige cities to process and filtrate their sewage.[51]

More generally, by 1900 we can observe a growing cultural appreciation of the value of natural landscapes, of monuments not only of architecture but also of nature. In 1904 the 'Bund Heimatschutz', the Society for the Protection of the Homeland, was established and engaged in campaigns for the protection of certain endangered landscapes.[52] One of its protagonists, Paul Schultze-Naumburg, an architect, published a series of well-illustrated books with scathing attacks on the ongoing degradation and debasing of countryside and the built environment.[53] In 1906 the Prussian government set up a department for the protection of natural monuments, the first institutionalisation of that field of policy in Europe.[54] And in 1911 the Lueneburg Heide was declared a natural monument, an area which, ironically, initially resulted from environmental degradation due to deforestation and overgrazing since the Middle Ages.

By the early twentieth century pollution of water and air had become political issues on various levels and there were debates over whether an Imperial Water law would be necessary. The powerful and well-funded lobby of the chemical industry however succeeded in preventing such a law. But the debates show a raised awareness that there were problems to be dealt with. Some German environmental historians claim that the movements for natural protection and 'Heimatschutz' (protection of landscape and natural habitats), together with parts of the hygiene movement, had reached a point just before the First World War where an effective policy of environmental protection could and might have been initiated.[55] Joachim Radkau claims that '… the single most important factor in dealing with environmental degradation in the 19th and early 20th century, in Germany as well as in Western Europe, has been the hygiene movement'.[56] Radkau underscores the very wide focus of the hygiene movement, which was concerned not only with cleanliness of the individual but also with the health of society as a whole. The defeat of Germany in the First World War and the following traumatic experiences of inflation, occupation and depression abruptly terminated such trajectories towards more effective environmental protection by state legislation and brought about a revised set of political priorities: the preservation of a natural environment now had – on the whole – to take backstage to economic growth and higher rates of employment.[57]

Let me close with a quote from one of the main promoters of the hygiene movement in the early twentieth century, the industrialist Karl Lingner, specialising in the production of a deodorising mouthwash-liquid 'Odol', who sponsored the International Hygiene Exhibition in Dresden in 1911, from which the German Hygiene Museum resulted. On the opening of a hygiene exhibition in Darmstadt 1912 he explained the social purposes of hygiene:

> Hygiene should not be a secret science. She is a free and cheerful science, taking a high flight towards a beautiful, sublime goal. Sun and Light, Air and Water are her allies, fresh and happy activity of the freely-developed body, to create cheerful and healthy human beings is her wonderful aim.[58]

Notes

1. T. Weyl (ed.) (1893–1901). *Handbuch der Hygiene* [Handbook of Hygiene], 10 vols (Jena: Gustav Fischer).
2. T. Weyl (1893). *Handbuch der Hygiene*, vol. 1, Vorwort, I (transl. D.S.).
3. The fact that Weyl addresses issues of Prussian policy and government portfolios in his preface refers to the constitutional structure of the German Empire: due to its character as a federation of princes and territories issues of domestic policy (to which health policy also belonged) were not under the direct responsibility of the Empire but rather of the governments of the several states such as Prussia. Since Prussia was by far the greatest power, comprising c. two thirds of population and of territory of the German Empire, it usually also played a leading role among the other states. See H. P. Ullmann (1995). *Das Deutsche Kaiserreich 1871–1918* [The German Empire 1871–1918] (Frankfurt am Main: Suhrkamp), pp. 31–3. Suggesting a 'ministry of health' might have recalled earlier comparable demands voiced by Rudolf Virchow in the context of the 1848 revolution and his progressive medical journal *Medicinische Reform*. See D. Schott (1999). '"Die Medizin ist eine soziale Wissenschaft": Rudolf Virchow und die "Medicinische Reform" in der Revolution 1848/49' [,Medicine is a social science'. Rudolf Virchow and the "Medizinische Reform" in the Revolution of 1848/49,], in Geschichtswerkstatt (ed.). *Die Revolution hat Konjunktur. Soziale Bewegungen, Alltag und Politik in der Revolution von 1848/49* [The Revolution is in Fashion.

Social Movements, Everyday Life and Politics in the Revolution of 1848/49] (Münster: Westfälisches Dampfboot), pp. 87–108.

4. See on Max von Pettenkofer and his pivotal role for the rise of hygiene E. Jahn (1994). *Die Cholera in Medizin und Pharmazie. Im Zeitalter des Hygienikers Max von Pettenkofer* [Cholera in Medicine and Pharmacy. The Age of the Hygienist Max von Pettenkofer] (Stuttgart: Steiner); A. Labisch (1992). *Homo Hygienicus. Gesundheit und Medizin in der Neuzeit* [Homo Hygienicus. Health and Medicine in the Modern Era] (Frankfurt am Main/ New York: Campus), pp. 120–32; on scientification in medicine in general see T. Nipperdey (1993). *Deutsche Geschichte 1866–1918. Vol. 1. Arbeitswelt und Bürgergeist,* [German History 1866–1918. Vol. 1. World of Work and Middle-Class Mentality] (München: Beck, 3rd edn), pp. 150–66; A. Hardy (2005). *Ärzte, Ingenieure und städtische Gesundheit. Medizinische Theorien in der Hygienebewegung des 19. Jahrhunderts* [Physicians, Engineers and Urban Health. Medical Theories in the Hygiene Movement of the 19th Century] (Frankfurt am Main/New York: Campus), pp. 201–06.

5. For the evolution of public health since Chadwick see C. Hamlin (1998). *Public Health and Social Justice in the Age of Chadwick* (Cambridge: Cambridge University Press).

6. See Pettenkofer's elaboration of the causes of cholera and the consequences thereof for public hygiene, R. J. Evans (1990). *Tod in Hamburg. Stadt, Gesellschaft und Politik in den Cholerajahren 1830–1910* [Death in Hamburg. City, Society and Politics in the Years of Cholera, 1830–1910] (Reinbek bei Hamburg: Rowohlt); Jahn *Die Cholera*, pp. 37–47.

7. See on sanitary engineering in German cities J. v. Simson (1983). *Kanalisation und Städtehygiene im 19. Jahrhundert* [Sewerage and Urban Hygiene in the 19th Century] (Düsseldorf: VDI); P. Münch (1993). *Stadthygiene im 19. und 20. Jahrhundert. Die Wasserversorgung, Abwasser- und Abfallbeseitigung unter besonderer Berücksichtigung Münchens* [Urban Hygiene in the 19th and 20th Century. Water Provision, Sewage and Waste Disposal under specific attention to Munich] (Göttingen: Vandenhoek & Rupprecht).

8. See on the significance of 'Bauhygiene', Hardy, *Ärzte, Ingenieure und städtische Gesundheit*, pp. 243–72.

9. See J. Rodriguez-Lores (1985). 'Stadthygiene und Städtebau: Zur Dialektik von Ordnung und Unordnung in den Auseinandersetzungen des Deutschen Vereins für Öffentliche Gesundheitspflege 1868–1901', [Urban Hygiene and Urbanism. On the Dialectics of Order and Disorder in the Disputes of the German society for Public Health 1868–1901], in: J. Rodriguez-Lores and G. Fehl (eds). *Städtebaureform 1865–1900. Vom Licht, Luft und Ordnung in der Stadt der Gründerzeit,* [Reform of Urban planning. On Light, Air and Order in the City of the Foundaion Period] vol. I (Hamburg: Christians), pp.19–58; J. Büschenfeld (1997). *Flüsse und Kloaken. Umweltfragen im Zeitalter der Industrialisierung (1870–1918)* [Rivers and Cesspools. Environmental Issues in the Age of Industrialization (1870–1918)] (Stuttgart: Klett-Cotta), pp. 51–63; Labisch *Homo Hygienicus*, pp. 127–32.

10. See on the co-existence of both etiological theories of disease causation for a considerable number of years Hardy, *Ärzte, Ingenieure und städtische Gesundheit*, pp. 345–72, who argues, that in terms of practical preventive action the germ theory did not cause major changes (p. 345); M. Worboys (2000). *Spreading*

Germs. Disease Theories and Medical Practice in Britain, 1865–1900 (Cambridge: Cambridge University Press), argues that '… it was only after 1895 that bacteriologist played a major part in public health', p. 234 and proposes a much more gradual, less clear-cut transition between the two approaches, particularly ch. 7, pp. 234–76.

11. This is particularly the argument proposed by Richard Evans in his magisterial study of the Hamburg cholera.
12. Hardy, *Ärzte, Ingenieure und städtische Gesundheit,* p. 371, uses this term by Tom Beauchamp to characterize the resolution of the debate at the Marburg assembly of the German Society for Public Health in 1894. At this meeting proponents of both the sanitarians and the bacteriologists (Kerschensteiner for the sanitarians and Gaffky for the bacteriologists) presented their views in such manner as to make it possible to incorporate the valid and proven aspects of the other theory. However, the causative role of the germ was generally accepted while Robert Koch also admitted the relevance of secondary causes, such as local, temporal and individual circumstances; pp. 367–72.
13. The caption of the first article reads 'Organisation der öffentlichen Gesundheitspflege in den Kulturstaaten', [Organization of public Health in cultured States] by Prof. Finkelnburg. 'Kulturstaaten' usually denoted the more advanced European (and North American) states; T. Weyl (1893). 'Vorwort', in idem (ed.). *Handbuch der Hygiene,* vol. I., (Jena: Gustav Fischer), pp. i–v, quote iv.
14. See J. Grützig/H. Mehlhorn (2005). *Expeditionen ins Reich der Seuchen: medizinische Himmelfahrtskommandos der deutschen Kaiser- und Kolonialzeit* [Expeditions into the Realm of Epidemics: Medical Death Missions of the German Empire and Colonial Period] (Heidelberg: Elsevier). Those years also witnessed the establishment of institutes for tropical medicine, for instance in Hamburg, where due to shipping contact with tropical diseases was most likely. The initiative came from Bernhard Nocht, who had become port physician in Hamburg in 1893. On 1 October 1900 the 'Institut für Schiff- und Tropenkrankheiten' (Institute for Ship and Tropical diseases) was established in Hamburg. Bernhard-Nocht-Institut (2009) http://www.bni-hamburg.de/, accessed 30 August 2009.
15. See Stübben O. Karnau (1996). *Hermann Josef Stübben. Städtebau 1876–1930,* [Hermann Josef Stuebben. Urban Planning 1876–1930], (Braunschweig/ Wiesbaden: Vieweg); B. Ladd (1990). *Urban Planning and Civic Order in Germany, 1860–1914,* (Cambridge, MA., London: Harvard University Press).
16. J. Stübben (1890). *Der Städtebau,* [Urban Planning] (Reprint: Braunschweig: Vieweg, 1980).
17. 'Children's Protection' ('Kinderschutz') dealt with specific health problems and health requirements of children such as child-specific demography, perinatal problems, feeding, infant childcare in creches, typical childhood diseases and their prevention, public health measures to promote the health of children. It also included social and psychological aspects, such as the attention to neglect and abuse without immediate health consequence, measures against child begging, child labour and relevant legislation. 'School hygiene' dealt with the construction and equipment of school buildings according to hygienic principles, particularly issues of building materials, exposure to sun, ventilation, equipment by furniture, heating, lavatories and toilets, but also the structuring

of school in terms of time-tabling, the numbers of pupils in a classroom, the frequency and length of breaks etc. See tables of contents of *Handbuch* No. 7.

18. Weyl, 'Vorwort zu "Soziale Hygiene"'. *Handbuch der Hygiene*, 4. Supplementband, S. vii. (transl. D.S.)

19. See Labisch, *Homo Hygienicus*, p. 120–3 on Pettenkofer's notion of Hygiene and p. 146f on conditional hygiene.

20. Labisch, *Homo Hygienicus*, p. 164 refers to contemporary publications by O. Naegeli. See on tuberculosis Flurin Condrau (2000). *Lungenheilanstalt und Patientenschicksal. Sozialgeschichte der Tuberkulose in Deutschland und England im späten 19. und frühen 20. Jahrhundert*, [Lung Rehabilitation Clinic and Patients' Fate. Social History of Tuberculosis in Germany and England of late 19th and early 20th Century] (Göttingen: Vandenhoek & Ruprecht).

21. See for a critical assessment U. Frevert (1985). '*Fürsorgliche Belagerung. Hygienebewegung und Arbeiterfrauen im 19. Und 20. Jahrhundert*', [Benevolent Siege. Hygiene Movement and Workers Women in the 19th and 20th Century], *Geschichte und Gesellschaft*, 11, pp. 420–46.

22. Condrau, *Lungenheilanstalt und Patientenschicksal*, pp. 104–111.

23. Labisch, *Homo Hygienicus*, p. 159.

24. Labisch, *Homo Hygienicus*, p. 148.

25. See V. Berghan (2003). *Das Kaiserreich 1871–1914. Industriegesellschaft, bürgerliche Kultur und autoritärer Staat* [The Empire 1871–1914. Industrial Society, Middle-class Culture and Authoritarian State] (Stuttgart: Klett-Cotta), pp. 58–9, 218–20, 282–4, 389–93; Ullmann *Das Deutsche Kaiserreich*, p. 138–45.

26. Berghan, *Das Kaiserreich*, pp. 312–3.

27. The share of Reichstag members from the SPD was however always below their proportional share of votes, due – apart from the second ballot system – to the fact that electoral districts, initially cut to a size of approximately 100,000 voters each, were not changed despite massive shifts in population due to industrialization and internal migration. In the later period of the Empire this tended strongly to favour conservative agrarian interests and disadvantage parties with a largely urban and industrial electorate such as the SPD; see Berghan, *Das Kaiserreich*, p. 321–2.

28. See on regional patterns of industrialization H. Kiesewetter (2004). *Industrielle Revolution in Deutschland. Regionen als Wachstumsmotoren* [Industrial Revolution in Germany. Regions as Engines of Economic Growth] (Stuttgart: Steiner); Berghan, *Das Kaiserreich*, pp. 63–4.

29. Far-reaching confidence in the beneficial effects of progress via technology is highlighted in statements made by the Frankfurt Mayor Adickes and other dignitaries at the opening of the International Electricity Exhibition in Frankfurt, May 1891, which brought the public introduction of three-phase current, a milestone in the technological history of electricity and the hope of eliminating steam engines from urban areas and substituting them with electric power imported from 'clean' sources such as hydropower far away; see D. Schott (1999). 'Das Zeitalter der Elektrizität: Intentionen – Visionen – Realitäten', [The Age of Electricity: Intentions – Visions – Realities] in *Jahrbuch für Wirtschaftsgeschichte*, 2, pp. 31–49.

30. A notable and influential forum for such 'progressive' debate and critique was the 'Verein für Sozialpolitik', set up in 1872 to discuss issues of social policy, conduct enquiries on the situation of craftsmen or housing conditions, and

develop proposals for legislation on the imperial level; see Nipperdey, *Deutsche Geschichte*, pp. 666–7. The men (almost exclusively) engaged in this and other societies were no longer 'liberals' in the classical sense, although most of them would have voted for liberal parties at elections. But they believed in the necessity and legitimacy of state intervention to correct and redress the apparent evils of unfettered capitalism. Particularly prominent prototypes of such 'progressives' were important mayors of large cities, e.g., Franz Adickes who developed a comprehensive municipal policy with a pronounced social agenda. See W. Klötzer (1981). 'Franz Adickes. Frankfurter Oberbürgermeister 1891–1912', [Franz Adickes. Mayor of Frankfurt 1891–1912] in Klaus Schwabe (ed.). *Oberbürgermeister*: [Mayors] *Büdinger Forschungen zur Sozialgeschichte* (Boppard am Rhein: Boldt), pp. 39–56.

31. On municipal activities in health and hygiene, Nipperdey, *Deutsche Geschichte*, p. 160f.

32. See K. Bergmann (1970). *Agrarromantik und Großstadtfeindschaft* [Agrarian Romanticism and Hostility towards the Metropolis] (Meisenheim am Glan: Hain); C. Engeli (1999). 'Die Großstadt um 1900. Wahrnehmung und Wirkungen in Literatur, Kunst, Wissenschaft und Politik', [The Metropolis around 1900: Perceptions and Effects in Literature, Art, Science and Politics] in C. Zimmermann and J. Reulecke (eds). *Die Stadt als Moloch? Das Land als Kraftquell? Wahrnehmungen und Wirkungen der Großstädte um 1900* [The City as a Moloch? The Country as a Source of Strength? Perceptions and Effects of Large Cities around 1900] (Basel, Boston, Berlin: Birkhäuser), pp. 21–51; A. Lees (1992). 'Das Denken über die Großstadt um 1900. Deutsche Stellungnahmen zum urbanen Lebensraum im internationalen Vergleich', [Thinking on the Metropolis around 1900. German Voices on the Urban Space in International Comparison] in *Berichte zur Wissenschaftsgeschichte*, 15, pp. 139–50.

33. See on general aspects of German urbanization J. Reulecke (1985). *Geschichte der Urbanisierung in Deutschland* [History of Urbanization in Germany] (Frankfurt am Main: Suhrkamp); C. Zimmermann (1992). *Die Zeit der Metropolen. Urbanisierung und Großstadtentwicklung* [The Age of the Metropoles. Urbanization and the Development of Metropoles] (Frankfurt am Main: Fischer).

34. Nipperdey sees a general feeling of crisis and critique of classical 'liberal' values among German intellectuals of turn of the century: *Deutsche Geschichte*, p. 591. The feeling of crisis gave expression to a yearning for 'wholeness' and synthesis against specialisation and quantitative growth. Thomas Rohkrämer posits that 'Zivilisationskritik' was aiming for a different kind of modernity: T. Rohkrämer (1999). *Eine andere Moderne? Zivilisationskritik, Natur und Technik in Deutschland 1880–1933* [A Different Modernity? Critique of Civilisation, Nature and Technology in Germany 1880–1933] (Paderborn: Schöningh).

35. E. Barlösius (1997). *Naturgemäße Lebensführung. Zur Geschichte der Lebensreform um die Jahrhundertwende* [Living according to Nature. On the History of Life Reform around the Turn of the Century] (Frankfurt am Main: Campus); K. Buchholz, R. Latocha, H. Peckmann and K. Wolbert (eds) (2001). *Die Lebensreform. Entwürfe zur Neugestaltung von Leben und Kunst um 1900*, [Life Reform. Projects on the Reshaping of Life and Art around 1900] (Darmstadt: Häusser).

36. U. Linse (ed.) (1983). *Zurück, o Mensch, zur Mutter Erde. Landkommunen in Deutschland 1890–1933* [Back, oh Man, to Mother Earth: Agrarian Communes in Germany 1890–1933] (München: dtv).

37. On the popularity of the Garden city concept in continental Europe F. Bollerey (ed.) (1990). *Im Grünen wohnen – im Blauen planen: ein Lesebuch zur Gartenstadt mit Beiträgen und Zeitdokumenten,* [Living in the Green – Planning in the Blue: A Reader on the Garden City with Essays and Sources] (Hamburg: Christians); A. Schollmeier (1990). *Gartenstädte in Deutschland: ihre Geschichte, städtebauliche Entwicklung und Architektur zu Beginn des 20. Jahrhunderts* [Garden Cities in Germany: Their History, urban Development and Architecture at the Beginning of the 20th Century] (Münster: Lit); H. Meller (2001). *European cities, 1890–1930s: History, Culture and the Built Environment* (Chichester: Wiley).

38. Reulecke *Geschichte der Urbanisierung,* p. 120f; W. Krabbe (1989). *Die deutsche Stadt im 19. und 20. Jahrhundert* [The German City in the 19th and 20th Century] (Göttingen: Vandenhoek & Ruprecht), pp. 139–42; W. Hofmann (1981). 'Oberbürgermeister als politische Elite im Wilhelminischen Reich und in der Weimarer Republik', [Mayors as Political Elite in the Wilhelmine Empire and the Weimar Republic] in: K. Schwabe (ed.). *Oberbürgermeister:* [Mayors] *Büdinger Forschungen zur Sozialgeschichte* (Boppard am Rhein: Boldt), pp. 17–38, 22.

39. This transition in German urban history is usually termed 'Leistungsverwaltung'; see J. Reulecke (1995). 'Einleitung', in idem (ed.). *Die Stadt als Dienstleistungszentrum. Beiträge zur Geschichte der 'Sozialstadt' in Deutschland im 19. und frühen 20. Jahrhundert* [The City as a Service Centre. Contributions on the History of the 'Social City' in Germany in the 19th and early 20th Century] (St. Katharinen: Scripta Mercaturae), pp. 1–17; see also A. Lees and L. H. Lees (2007). *Cities and the Making of Modern Europe. 1750–1914* (Cambridge: Cambridge University Press).

40. On the significance of municipalisation of gas, water etc., W. Krabbe (1990). 'Städtische Wirtschaftsbetriebe im Zeichen des "Munizipalsozialismus": Die Anfänge der Gas-und Elektrizitätswerke im 19. und frühen 20. Jahrhundert', [Urban Utilities under the auspices of ‚Municipal Socialism': The Beginnings of Gas and Electricity Works in the 19th and early 20th Century] in H. Blotevogel (ed.). *Kommunale Leistungsverwaltung und Stadtentwicklung vom Vormärz bis zur Weimarer Republik* [Municipal Service Administration and Urban Development from the Pre-1848 Period to the Weimar Republic] (Köln, Wien: Böhlau), pp. 117–35. E. P. Hennock (2000). 'The urban sanitary movement in England and Germany, 1838–1914: a comparison', *Continuity and Change,* 15, pp. 269–96.

41. On municipal health policy see J. Reulecke and A. Gräfin zu Castell Rüdenhausen (eds) (1991). *Stadt und Gesundheit. Zum Wandel von 'Volksgesundheit' und kommunaler Gesundheitspolitik im 19. und frühen 20. Jahrhundert* [City and Health. On the Change of ‚People's Health' and municipal Health Policy in the 19th and early 20th Century] (Stuttgart: Steiner); T. Bauer, H. Drummer and L. Krämer (1992). *Vom 'stede arzt' zum Stadtgesundheitsamt : die Geschichte des öffentlichen Gesundheitswesens in Frankfurt am Main* [From the „stede arzt" to Municipal Health Office: History of Public Health in Frankfurt am Main] (Frankfurt am Main: Kramer);

J. Reulecke (1996). *'Gesundheitsfür – und –vorsorge in den deutschen Städten seit dem 19. Jahrhundert'*, [Health Provision and Health Care in German Cities since the 19th Century] in D. Machule (ed.). *Macht Stadt krank? Vom Umgang mit Gesundheit und Krankheit* [Does the City Make Sick? About Dealing with Health and Sickness] (Hamburg: Dölling und Galitz), pp. 70–83.

42. See also Hardy, *Ärzte, Ingenieure und städtische Gesundheit* on the range and breadth of this movement.

43. Data on professional background, residence and age were compiled by my research student Astrid Gernhardt.

44. See Berghan, *Das Kaiserreich*, p. 149 on this political generation, referring back to reflections by Karl Mannheim. See also M. Doerry (1985). *Übergangsmenschen. Die Mentalität der Wilhelminer und die Krise des Kaiserreichs* [People of Transition. The Mentality of Wilhelmines and the Crisis of the Empire] (Weinheim: Juventa). Another prominent representative of the Wilhelmine generation was Max Weber (b. 1860) who in his famous inaugural address of Freiburg gave very vivid expression to this sense of belonging to an epigonal generation.

45. Agnes Bluhm (b. 1862) first trained as a teacher, then studied medicine in Zurich and trained as gynaecologist in Munich and Vienna. She established a medical practice in Berlin in 1890, which she ran until 1905 when she had to abandon practical medicine due to hearing problems. She contributed an article to vol. 8 of the *Handbuch* on hygienic care for female workers and their children. After 1918 she was a researcher at Kaiser-Wilhelm-Institute for Biology in Berlin, where she worked during the National Socialist period on issues of hereditary diseases and racial and social hygiene. Died 1944 in Beelitz, see *Deutsche Biographische Enzyklopädie*, vol. 1.

46. *Deutsche Biographische Enzyklopädie*, vol. 10, p. 467.

47. In Stuttgart, a city with a high degree of skilled labour, the weekly wage stood at 22 Marks, by 1900 the daily wage would be 3,66 Marks, see Berghan, *Das Kaiserreich*, p. 51.

48. The volumes of the *Handbuch* did not constitute individual books. Each comprised two (or more) 'compartments' ('Abteilung' or 'Lieferung') which were delivered and could be purchased on their own.

49. L. Burgerstein and A. Netolitzky (1895). 'Vorwort', in: T. Weyl (ed.). *Handbuch der Hygiene*, vol. VII, 1st compartment (Jena: Gustav Fischer), pp. ix–x, here x.

50. On river pollution and the debate over whether cities were allowed to discharge their sewage into rivers, Büschenfeld *Flüsse und Kloaken*, pp. 265–75.

51. See ibid.; D. Schott (2000). 'Remodeling "Father Rhine": The case of Mannheim 1825–1914', in S. Anderson and B. H. Tabb (eds). *Water, Culture and Politics in Germany and the American West* (New York: Lang), pp. 203–25.

52. J. Radkau (2000). *Natur und Macht. Eine Weltgeschichte der Umwelt* [Nature and Power. A World History of the Environment] (München: Beck) pp. 265, 270ff. Radkau underscores that the 'Bund Heimatschutz', pursuing above all an aesthetic approach, was most successful in protecting individual monuments of nature and culture; its primary interest was devoted to aesthetic values of landscapes rather than their ecological functions.

53. Paul Schultze-Naumburg (1907). *Kulturarbeiten*, [Cultural Works] *Vol. 1* (München: Callwey).

54. See Radkau, *Natur und Macht*, p. 265.

55. Frank Uekötter terms the Imperial Period as the 'Sattelzeit' of environmental history and emphasises that environmental debate in imperial Germany had reached an intensity and breadth which remained unparalleled until long after 1945; F. Uekötter (2007). *Umweltgeschichte im 19. und 20. Jahrhundert* [Environmental History in the 19th and 20th Century] (München: Oldenbourg), p. 22.

56. Radkau, *Natur und Macht*, p. 280, translation DS.

57. Uekötter, *Umweltgeschichte*, p. 23.

58. Karl Lingner, at the opening of the Darmstadt exhibition 'Der Mensch', 9 August 1912, quoted in *Darmstädter Tagblatt*, 10 August 1912.

5

Leagues of Sunshine: Sunlight, Health and the Environment

Simon Carter

5.1 Introduction

This chapter explores how, during the 1920s and 1930s, a variety of forces came into play to weave sunlight, as a giver of health, into the fabric of social environments. The use of sunlight, both natural and artificial, was already well-established as a therapy that had been prevalent in Europe since the late nineteenth century.[1] The growth of sunlight therapy, however, expanded greatly in the early part of the twentieth century, with heliotherapy (natural sunlight) being used to treat tuberculosis and actinotherapy (via sunlamps) deployed to combat rickets.[2] These therapeutic applications, applied in sanatoria and clinics, helped establish an association between sunlight and health. The growth in the use of such sunlight therapies was partially based on the idea of 'nature' being curative of the diseased body.[3] However, this idea of nature as curative and enhancing health was also being found in non-clinical developments in this period and these helped to produce equivalence between health and the desire to introduce sunlight into the environments used for living and working. It is these developments that this chapter will consider.

The establishment of sunlight as a source of health was in contrast to how the sun's rays had been commonly regarded. For example at the end of the nineteenth century solarisation was regarded as dangerous, with bodily isolation from the sun being the prescriptive norm. In particular exposure was commonly regarded in one of two ways: first, for men (particularly those serving the needs of empire and stationed overseas) the sun was seen as a source of danger leading to a variety of ills and even death;[4,5,6] and, second, for women (especially those with aristocratic aspirations) sun exposure was thought to lead to discomfort and be a marker of low social status.[7]

5.2 Scouts, campers and pastoralism

Be this as it may, in the first half of the twentieth century it was possible to observe a nexus developing around sunlight linking three broad sets of ideas – namely morality, health and nature. A healthy body could be produced by physical outdoor activity and this would in turn produce a wholesome and virtuous mind. Such a body exposed to the sun's rays would become brown, bronzed or tanned. The darkened body, tanned by the rays of the sun and exposure to the elements of nature was the most prominent indicator that an individual had engaged in such worthy activities. In many ways this discourse of the suntan was ultimately grounded in an idea of nature that was pure, uncomplicated and innocent. The acquisition of a suntan could here almost be described as applied Rousseau — an idea that would articulate well with the 'late-nineteenth-century love of the primitive' and the 'noble savage'.[8] Vitality could be gained by mind and body through a return to the innocence of nature.

The idea of nature as a restorative force was at the centre of a number of social movements, which were themselves forging links between ideals of health and 'worthy' outdoor activities conducted in sunshine. These movements, often aimed at young men, were partly fuelled by worries about the growing size of urban industrial conurbations and the possible drift, by a large section of the society, into physical or moral degeneracy.[9,10] The perceived susceptibility of young men to 'moral degeneracy' was a focus of particular anxieties about the possible negative effects that any unchecked sexual awakening may have as the boys entered adulthood. One solution to this problem was thought to be the pursuit of 'muscular Christianity': the idea, established in British public schools, that 'healthy' manliness could be obtained through disciplined and codified physical activity. The notion that informed these activities was the perceived link between the values of physical health, manliness and morality.[11] These values were often contrasted to the moral dangers associated with the enfeebling environment of the modern city and led to calls for a new stress on the simpler 'spartan life' lived within 'nature'. If boys could be removed from the urban environments and immersed in 'nature' then such surroundings would lead to a situation where the 'emphasis was likely to be on energetic action rather than unhealthy reflection'.[12] It was in this way that the suntan served as a physical marker of differentiation from the unhealthy, amoral and pallid bodies found lurking in the darker environments of the city.

Two social movements of this period sought to introduce these beliefs to a wider group of young males: the Boy's Brigade (established in 1872) and the Scout movement (Baden-Powell published his *Scouting for Boys* in 1908). Both these movements put a great stress on the health benefits to be gained from open air physical training for boys. The Scout movement in particular emphasised the benefits of an outdoor life, away from the city, for the young man. The rural location of the camp was thought to promote a link between 'the healing and regenerative power of nature and the development of the whole personality'.[13] Further, within the Scouting literature, there was also a link between health and the acquisition of a suntan. Here we can see how the tan became an indicator or marker that a young man was engaging in healthy outdoor activity. The tan represented a visible link between the body and idealised concepts of nature and masculinity:

> He is a hefty Rover Scout, about seventeen years of age, that is a fellow training to be a man... In addition to his load he carries a more important thing – a happy smile on his weather-tanned face. Altogether a healthy, cheery young back-woodsman. Yet this chap is a 'Townie', but one who has made himself a Man.[14]

The temporary return to nature was an idea that not only informed the Scout movement but was also partly responsible for a revision of popular opinion about camping. Up until the end of the nineteenth century the word 'camp' was primarily associated with the temporary dwellings used by either the military. A number of factors had begun to coalesce and change the position of camping within popular imagination, for example: a romantic literature extolling the virtues of an heroic camp life away from the dull and predictable city; and reports in the daily press and Sunday school slide shows about the exploits of missionaries and explorers and their campsite living. In addition there existed an established industry manufacturing a wide variety of camping accoutrements to serve the needs of the military: 'camp chairs, camp beds, camp stoves, camp kettles for the camp fire, Camp Coffee and Camp Matches'.[15] This industry was easily able to re-purpose their commodities towards meeting the needs of the hobbyist camper. As the idea of leisure camping took hold an emerging literature providing practical help for the would-be camper. For example, Thomas Hiram Holding took a proselytising role with regard to camping and played a key role in the formation of the *Association of Cycle Campers* in 1901 and the *National Camping Club* in 1906. In establishing camping as a leisure activity Holding wrote the

Cycle and Camp in 1907 and the *Camper's Handbook* in 1908. In these works we find frequent references to the value of the simple life in proximity to fresh air and the benefits of sunlight exposure. Indeed the original cover of *Cycle and Camp* featured a woodcut depicting a couple siting outside a tent under a large, stylised sun. When he wrote of the rewards of camping Holding explicitly mentions the suntan – thus in the following extract a family camp is described:

> Here the family and their servants were spending a 'savage' holiday. . . . They were having a delightful time. The *brown* limbs of the children, the *bronzed* faces of the parents and the grown of branches of the family. . . . At the end of the month they were not tired, but were counting with regret the remaining days in camp.[16]

Both the Scouting movement and the increasing popularity of leisure camping at the turn of the twentieth century were themselves developing in parallel with wider popular movements of the time. These could be loosely associated with the tradition of pastoralism – the 'image of lost rural bliss and to an affinity with Nature'.[17] This tradition, with its nostalgic overtones, sought to recreate a purer and more natural way of life. It was a 'fantasy of gentle scenery, of mellow farms and villages and of beautiful people filled with love; it is a world of perfection and harmony, of the Garden of Eden before the Fall'.[18] This romantic attraction to nature was something that had been advancing for most of the nineteenth century, as superstition gave way to scientific discovery, nature in turn became something to 'admire and discover'. This connected to wider ideals of romanticism in circulation at this time – a return to a simpler life that was nearer to nature, but one that was 'a synthesis of intellect and emotion, of rationality and imagination'.[19] This reification of nature was also in opposition to the perceived distortion caused by modern urban life and was expressed by writers in this era such as William Morris[20,21] and Edward Carpenter. Indeed Carpenter, commenting on Victorian family life, said 'Plain food, the open air, the hardness of the sun and wind, are things practically unobtainable in a complex ménage... No individual or class can travel far from the native life of the race without becoming shrivelled, corrupt, diseased...'.[22]

5.3 Sunlight and social hygiene

Yet the interest in the moral and health benefits of nature, the outdoors and sunlight did not occur in isolation. The emergence of ideas about the deleterious effects of the city and the benefits of sunlight and nature cannot be disassociated from social changes that were taking place in this period, particularly in ideas about medicine and public health. Between the late-nineteenth and mid-twentieth century it can be said that medicine was in state of flux, with major transformations taking place.[23] In particular, and at the centre of this shift in the focus, were concerns with efficiency in which a diversity of interests came to be subordinated to serving the 'emerging corporate and national interests that demanded the creation of an increasingly unified system of mass health care'.[24] These changes also produced a repositioning of the social relations of medicine with the individualisation of therapeutic care[25] and patronage, typical of private practice, giving way to an ethos of professional management underpinned by the use of scientific methods (e.g. both scientific management and the application of laboratory sciences). However in this process, and partly in response to the increasing biomedical orientation towards clinical questions, there emerged a variety of diverse and different approaches.[26] It is difficult to map out these alternatives precisely as they came from a variety of intellectual traditions and practices, both external and internal to medicine. Some drew on specific or partial elements of the emerging biomedical sciences and the developing orthodoxy (e.g. Darwinism and evolutionary theory) while others expressed an outright opposition to the way medicine was heading – many were '*bricoleurs,* putting together distinct packages for specific uses in varying contexts'.[27]

One such grouping was the social hygiene movement, which was itself a fusion of several nineteenth century concerns, most notably around sanitary reform and public health. As Rosen has indicated, throughout the nineteenth-century surveys and statistical movements in various European countries (especially Germany, Belgium and Britain) exposed how ill-health was patterned by both geography and social class.[28] When examining the great outbreaks of cholera and typhus in the nineteenth century, it soon became evident that there was a strong link between material measures of social class and mortality, with particularly high rates of infant and maternal mortality in the poorest areas of the newly emerging industrial cities. There was much speculation about the possible causes of these differential rates of ill health and some looked towards the social, environmental and material conditions of the poor. However, the habits and domestic management of the underprivileged were also blamed. It was a short step from this to looking for hereditary explanations for the ill health of the poor.[29] Indeed the fusion of evolutionary and sociological theories led to the common view that evolution may be going backwards.[30] Thus the final element of social hygiene was a strong eugenic component.

The core ingredients of the movement were never developed into a coherent theoretical model; rather they were a series of interlocking ideas to be drawn on in a piecemeal fashion that brought together a range of political positions. But the stress on hereditary arguments, about poverty and ill health, sat well with the connection drawn between national economic efficiency and the good health of the population within social hygiene. In short many of the ills of the new urban industrial conurbations could be blamed on the poor and what was seen as their inadequate breeding rather than any structural problems due to the distribution of resources within an industrialised economy.

Two health-reform organisations that had their foundations in the social hygiene movement are of relevance here and both shared a belief that the aspects of the environment could both, contribute to, but could also be deployed to combat degeneracy. First, the People's League of Health which was established in 1917 by Olga Nethersole after she became concerned with the unhygienic conditions of working-class homes – 'I knew that slums breed disease, moral and physical'.[31] Its aims were wide-ranging but included imperial regeneration, eugenics, improvement of young people's health and immigration control, along with a desire to improve the nation's housing, sanitation and nutrition. However, another central interest of the League was the promotion of sunlight as a way of improving hygiene. Thus clinicians who had promoted sunlight as a health therapy, such as Leonard Hill and Sir Henry Gauvain, were members of the League and addressed public meetings on the benefits of sunlight. At the first annual general meeting of the People's League of Health in 1922 the importance of sunlight was given a prominent place by Caleb Saleeby, another member of the League:

> Sunlight prevented certain complaints of children and...it was not only a great antiseptic, but was of great food value, and of value in banishing disease. The motto of the league should be – "Back to the light"... The restoration of sunlight to our children would be the greatest task of hygiene.[32]

In 1924 the League made representations to the Minister of Labour, Margaret Bondfield, at the House of Commons in order for the government to consider resolutions passed at the People's League of Health First International Conference. Both Hill, who had addressed the conference, and Saleeby spoke to the minister of the important benefits to be gained by sunlight exposure. Hill commented on the importance of sunlight being admitted to dwellings and Saleeby spoke on the need for smoke abatement to prevent winter sunlight being curtailed.[33]

The second society with an interest in promoting the use of sunlight was the New Health Society established by William Arbuthnot Lane in 1926. This group focused on improving the nutrition of the nation and reflected the specific theories of its founder – that many modern diseases were caused by poor diet and 'intestinal stasis'. Therefore, the Society concentrated its efforts on trying to educate the public about how to best obtain a proper diet. However, hereditary factors as well as diet were 'considered to influence the nation's health and both were considered to have been adversely affected by urban growth'.[34] As well as the concerns of nutrition and the negative effects of urban growth, the Society promoted the idea that sunlight exposure had a nutritional value and may be a valuable way to supplement a 'proper diet'. For example, the New Health Society was active in promoting the invention of 'Vitaglass' in the 1920s. Vitaglass was a form of glass, invented by Pilkington Brothers, which allowed ultraviolet rays to pass through unhindered (normal glass blocks ultraviolet rays). Thus in 1926 the Society hosted a press conference at the London Zoo to mark the experimental installation of Vitaglass in the animal houses. Dr Belfrage used the occasion to explain that the Society was putting a high value on the invention of this new glass; 'ultra violet consists of invisible rays...It had been proved that they had a stimulating effect on general growth, power of resistance to disease, and on the richness of blood... ordinary window glass... was quite un-transparent to the health giving ultra-violet rays'.[35] The installation of Vitaglass was recommended for all schools and hospitals, and advertising materials for the product claimed that it had already been installed in 200 schools and 300 hospitals.

The New Health Society also advocated the resettlement of rural areas by the urban masses or at least the provision of allotments and[36] gardens to the urban poor. This would accomplish several objectives simultaneously: it would allow the consumption of a natural diet produced directly from the land; it would allow continued exposure of the urban masses to sunlight and fresh air; and it would expose the masses to conditions of 'hardness' in which the fittest would excel and be 'naturally' selected. Underpinning this is the idea that these measures would cleanse society of 'the discontents and unemployables who clog the wheels of progress, create disharmony and foster revolution'.[37]

Just as the promotion of sunlight, by those in the social hygiene movement, became more commonplace, so too did the advocacy of sunlight exposure in the popular media. Throughout the 1920s *The Times* carried regular features of the use of sunlight (natural and artificial) for the treatment of various clinical conditions but on 22 May 1928 *The Times* carried a special edition entitled the *Sunlight and Health Number*. This special supplement was around 30 pages long and carried features on every aspect of sunlight and health, including stories on: heliotherapy, winter sunlight, schools in the open air, sunlight and home decoration, garden cities for sunlight, open-air culture in Germany, artificial sunlight apparatus and industrial hygiene. There was also an extended review of all the possible health resorts that people may visit to enjoy sunlight, including: the spas of Great Britain, the French Riviera, Spain, Switzerland, Algeria, Egypt, Vichy, Aix-Les-Bains and the Canary Islands. However the diversity of articles presented in this special issue go far beyond the purely health benefits of sunlight. Indeed some articles focus on forms of home decoration and architecture that may be deployed to complement the effects of sunlight. Moreover the features on health resorts were obviously aimed at the pleasure seeker as much as the invalid, with information about casinos, shopping and sports.

The editorial of the *Sunshine and Health Number*, however, retains a serious focus on the health benefits of the sun's rays and frames the use of sunlight as a 'new science'. This editorial first establishes the long history of sunlight therapy, with current usage being 'a rediscovery of knowledge which at one point of history was widely disseminated'. The editorial then goes on to establish the scientific credentials of sunlight treatment because it is 'based on observation and on study which belong peculiarly to the present stage of the evolution of science'. Finally the need for caution is stressed because 'sunlight treatment is medical treatment in the strictest sense of the term and ought to be given only by physicians who have devoted special study to it'.[38] Of course this warning, that the therapeutic use of sunlight should only be undertaken under medical supervision, frequently appeared in the medical literature as well. Thus Gauvain, a British tuberculosis specialist and sunlight advocate, writing in the *Journal of State Medicine*, spoke of the need for the need for careful monitoring of the body because sunlight was a 'tonic, and like all tonics must be wisely employed and not abused'.[39] In a similar way an article, in a 1929 copy of the *Journal of the American Medical Association*, compares the therapeutic use of sunlight to other types of radiation, '… sun rays, like x rays and radium, posses the property of doing harm, instead of the good which is desired of them… The universally tanned skin takes many weeks to attain'.[40]

The feature immediately following this opening editorial seeks to establish the general significance of the health benefits to be gained from sunlight by making much of the increased mortality rates during the winter months in Britain. A number of graphs and charts are reproduced showing mortality rates by month of the year plotted against the amount of sunshine. According to the author these demonstrate that the 'days of darkness are also the days of death and disease'.[41] Thus this feature seeks to position the presence of sunlight as a positive and health giving feature of the physical environment – as a natural and abundant resource that could potentially provide health benefits to all.

5.4 Caleb Saleeby, the Sunlight League and smoke abatement

We can begin to see how sunlight was being framed as an entity that may be of value in raising the general resistance of the body to disease – it was becoming a 'hygienic' factor for the promotion of good health – a tonic that may be of benefit to all. A number of factors were steadily producing the sun's rays as an environmental factor that may strengthen and invigorate the human body, and in the interwar years several additional forces came into play to further position sunlight as a vital environmental factor in the fabric of social worlds. We have just seen how the People's League and the New Health Society played an active role in campaigning about the benefits of sunlight for public health. However, their primary role was not the promotion of sunlight itself. Rather these social movements sought to frame the health benefits of sunlight exposure as one factor among many, such as diet or the reproductive control of the 'enfeebled', that might be able to stabilise a new 'healthy' social figuration. In this way the population would, via the application of social hygiene, become fit and strong. This was an interactive process in which sunlight needed stabilising as much as the population. This was because both the sun's rays and the population were unaccommodating allies: the sun's rays might be obscured by weather or by smoke pollution; there was a seasonal variation in the power of sunlight; houses and dwelling, especially in poor areas, did not admit sunshine easily; and the populace were resistant to removing their clothing to expose their bodies.

However, in the early 1920s, an organisation appeared that sought to promote actively the action of the sun's rays as a benefit in their own right and to materially 'domesticate' the sun's rays – the Sunlight League founded in 1924. The story of the League is intimately connected to the career of its founder Saleeby, a medical doctor who gave up his clinical career in 1901 soon after graduating from Edinburgh University. He instead focused work for private voluntary organisations, freelance journalism and other writing. In the first decades of the twentieth century he regularly wrote a column for *The New Statesman* under the pseudonym of 'Lens'. However, before this in 1901, a clinical encounter as a student physician in the newly established Royal Maternity Hospital was to become a formative experience. Here he attended the first patient to be admitted, a woman in labour who was also suffering from rickets. He wrote, with obvious anger, of the incident on a number of occasions in subsequent years:

> The first occupant of that bed, a little rickety woman from the sun starved smoky Leeds, as brave as a lioness, reached us too late and died after a caesarean section and her infant in my arms a few hours later. It was a tragic beginning for one of the best ideas in the long history of hospitals.[42]

The episode left a lasting impression on Saleeby and informed his later interests in both sunlight and eugenics. In his early career he sought to popularise eugenic philosophies but soon fundamental disagreements emerged with the officers of the Eugenics Education Society over his attacks on the 'better dead' school of eugenics, which he accused of discrediting the movement with their 'reactionary class prejudices'.[43] Indeed he argued that many of the ideas about 'natural selection' favoured by the wider eugenic movement were mistaken. For example, he addressed the falling rates of infant mortality and how, from a simple eugenic position, this should lead to increased rates of degeneracy but that this outcome was unsupported by empirical evidence:

> The natural elimination of the unfit has been, it would appear, in a very large degree and very suddenly suspended... Further, very striking and deplorable consequences should be evident at Bournvill and Letchworth [in New Zealand]... where infant mortality is now down around forty per thousand... under these conditions natural selection has no chance; nearly every baby lives; what a pitiful crowd of degenerates the adult population must be! But we all remember the Apollo-Hercules type whom New Zealand sent us in the war – the Scot in excelsis... perfect and all-significant contrasts to the typical product of the conditions in Glasgow and Dundee twenty years ago, when 'natural selection' really had a chance.[44]

The implication of his arguments was that the enemies of a healthy population were poverty and pollution. Saleeby had other notable differences with the wider eugenic movement, such as his continuing interest in promoting post-natal care and the gradual modification of his views away from the need for 'racial improvements' towards advocating policies to arrest, what he believed to be the quantitative and qualitative decline of existing populations. He thought that these issues were best dealt with by preventative medicine and education to combat what he called the 'racial poisons' of venereal disease, pollution, tobacco and alcohol. Saleeby was then an active part of the social hygiene movement and, as such, was a campaigning member of both Arbuthnot Lane's New Health Society and the People's League.[45] Despite his long association with the eugenic movement he was what may be described as a 'reformist' rather than 'conservative' eugenicist.

In the post-war years Saleeby began his advocacy of increased sunlight exposure as a strategy to be deployed in the project to strengthen the health of the general population. From the beginning he linked his support of sunlight to his hatred of smoke pollution within the city environment. For Saleeby, city smoke pollution was an enemy of good hygiene for two reasons: first, as a bronchial irritant, it was a threat to health in its own right; and second, smoke in the atmosphere blocked the sun's rays, especially ultraviolet rays, and thus prevented their 'health enhancing' effects from reaching the population below. The principal cause of smoke was the domestic use of coal to heat dwellings and Saleeby, while noting the ignored alternatives to coal, concluded that coal heating was 'designed to perpetuate the darkness, dirt, disease and death which shamefully distinguish our present cities'. Thus he argued that smoke led to what he termed the 'diseases of darkness', which included rickets and tuberculosis but also a range of other amorphous conditions: 'the darkness that can be smelt in cities promotes suicide, melancholia, drunkenness, depression of the mind as well as body'.[46]

In 1921 Saleeby visited August Rollier's sunlight clinic at Leysin, Switzerland. Rollier's sanatorium, *Beau Soleil*, had a reputation for being a clinic where those suffering from tuberculosis could sometimes go into remission and it was Rollier who had written one of the first textbooks on using sunlight as a medical therapy. This textbook was based on his experiences of administering sunlight treatments and was first translated into English in 1923, appearing under the title *Heliotherapy – with special consideration of surgical tuberculosis*. The book was reprinted several times in the 1920s. On his return to Britain, Saleeby subsequently wrote a series of articles in *The New Statesman* under the title of 'Modern Sun Worship'.[47,48,49] In these, together with his own book *Sunlight and Health* (1923), he described and positioned Rollier as an authoritative voice in modern medical practice. He was the 'High Priest of Modern-Sun Worship... and his temple is Leysin, in Switzerland'.[50] However for Saleeby the significance of the sun exposure involved in heliotherapy went far beyond clinical practice and the fight against tuberculosis:

> The therapeutic lessons of Leysin are of supreme importance, not at all in themselves, but because of the prophylactic lessons they teach, for our cities, homes, schools, workshops or mines, wherever they may be.[51]

Saleeby came to refer to the combination of smoke-abatement policies, together with prophylactic application of the sun's rays, as *helio-hygiene* – a concept that it was hoped would capture the idea of the health of urban populations being strengthened by exposure to sunlight under clear blue skies. The promotion of helio-hygiene very much underpinned the formation of Sunlight League in 1924, with Saleeby as the chairman and Queen Alexandra as the patron. One of the League's main activities was the regular publication of the journal *Sunlight: A Journal of Light and Truth*. Though this journal, the Sunlight League campaigned for a variety of causes such as mixed sunbathing, open-air sunlight schools and the general health benefits of sunlight. The first edition carried messages of support for the newly formed League from a number of eminent sources in politics, the sciences and the arts, among others these included: David Lloyd George, former prime minister and leader of the Liberal Party – 'the present generation has no more important and more hopeful task than to let sunlight into our towns'; Charles John Bond, chairman of the MRC Tuberculosis Committee and keen eugenicist – 'the true aim of the League is to spread knowledge... of the preventive influence of sunshine and fresh air in raising the resistance of the body to all forms of microbe infection'; the novelist H. Rider Haggard – 'the man who abolishes smoke in our great cities and lets more sunshine into their crowed streets will be one of the greatest benefactors of his age'; Arbuthnot Lane, the founder of the New Health Society – 'I am greatly interested in your Sunshine league'; the actor and theatre manager, Johnston Forbes-Robertson; and the director of the Royal Institution, W. H. Bragg.

The editorial foreword to the first edition warned that we were entering a second Industrial Revolution. Whereas the first Industrial

Revolution had been characterised by coal and steam, the second would be based on the new form of power – electricity – and on better forms of communication. This would present a chance for 'laying the foundation of a social system that shall satisfy all the aspirations of a great, intelligent and educated people'.[52] The editorial was thus couched in discourse of scientific progress leading to a new social arrangement. There were also mystical undertones discussing recent insights from physics: 'every individual is united with the totality of living beings by invisible bonds. All the living hold together, and all yield to a single impulsion'.[53] But the main emphasis was on forging new types of city life, where admission of sunlight was incorporated as an environmental element into a utopian view of progress:

> We declare war against the powers of darkness; smoke and slums must go...our new houses must be placed so as to receive the sun... We seek to multiply the sources of information and education to such an extent that... no man or woman shall be stricken with disease for lack of knowledge of the light that heals... it is ours to better the environment, and bring forth a fitter race for humanity's high purposes.[54]

Of course slum clearances and the provision of housing for the urban poor had been an issue since the nineteenth century. The public revelation, in contemporary accounts from this era, of the conditions endured by the urban poor caused widespread public scandal. For example the founder of the Salvation Army, William Booth, produced a famous exposure of the conditions under which the urban poor lived, *In Darkest England* (1890). Similarly James Hole's *The Homes of the Working Classes* (1866) described the oppressive environmental conditions created by nineteenth-century urbanisation. Publications, such as these provided a momentum for liberal reformers and one of the first attempts to address some of these issues was the Public Health Act of 1875, which attempted to clear away 'hidden slum courts' and brought into being the so called 'bye-law' streets where all the dwellings were required to have an exposed 'street front'.[55]

However after the First World War, house reform accelerated as the wartime coalition government of Lloyd George 'planned a post war housing policy, enshrined in the phrase "homes fit for heroes"'.[56] After the war it was acknowledged that private enterprise would not be able to supply both the quantities of new homes, and of a high enough quality. Thus the idea that central government finance could be used to subsidise local-authority building was established and enacted under the Housing Act of 1919. The problem of urban slums was further addressed by the Greenwood Act of 1930, which required local authorities to prepare slum-clearance plans together with adequate provision to rehouse those displaced from the cleared slums.[57] The building of new housing stock, on a scale and quality not seen before the war, was thus thought to be an opportunity to address the problem of city smoke by introducing forms of domestic heating that either used smokeless fuels or new technologies such as electricity or gas. Many local authorities rose to this challenge by building new homes that used electric or part electric heating. Others introduced gas central heating and gas cookers in place of old-fashioned solid-fuel powered kitchen ranges. However, as a rule the majority of homes still used smoky fuels for heating and cooking.[58]

It is against this background that the Sunlight League campaigned in its journal for smoke abatement in cities and the need to introduce sunlight into the houses of the urban masses. Thus in a paper on helio-hygiene Saleeby draws a comparison between the waterborne diseases of the nineteenth century city and the need for pure air and light in the twentieth century. He goes on to call the use of coal as a fuel for heating as a shameful waste of a national resource before suggesting a radical solution:

> If some barbarians... should burn the contents of the Bodleian Library, which are certainly combustible, we should reproach them for seeing nothing but fuel in such a priceless thesaurus... It was made by life, and is crowded with the incomparable products of the Master-Chemist... The answer to the question, How to Clean our Skies? Is therefore Cease to Burn Coal... There is no real remedy, but the radical one, to stop burning coal and distil it instead, like intelligent beings.[59]

Saleeby was himself a constant advocate of technological innovation as a solution to the smoke problem caused by domestic fuel. As the quote above implies, he urged that coal should be used as a raw material in the chemical industry (e.g. for the production of fertilisers). Heating should to be provided by electricity generated from renewable resources: 'as a student of urban hygiene, I am prepared to vote for water falls and hydro-electricity every time'.[60]

During the 1920s the League directly lobbied the then Minister of Health, Neville Chamberlain, on the Public Health (Smoke Abatement) Bill being laid before Parliament. A series of letters were reproduced in *Sunlight* and *The Times* documenting the exchange and laying out the League's position that exemptions contained in the proposed bill should be removed, especially those relating to coal fires in private dwellings. Chamberlain's reply stressed his own desire 'to do all that is reasonably possible to reduce the smoke evil' but also highlighted the danger of measures that may 'hinder recovery of trade' during a period of industrial depression.[61,62]

Of course the League was not alone in its campaign to limit the amount of smoke pollution in cities. Indeed, throughout the nineteenth century, commentators had remarked on the poor quality of air in British cities.[63] The desire for clean air condensed several anxieties around public health and illness but it also refers to more general rising concerns within the nineteenth century about public civility, visibility and the socio-sensual environment – the polluted darkness of city regions, such as the East End of London, was often linked to other social problems such as lawlessness and 'moral degeneracy'. Smoke pollution was ubiquitous – it imperilled all hygienic surfaces and passed across all boundaries and as such was of concern to many outside the Sunlight League.

Since the mid-ninetieth century a variety of anti-smoke societies had been active campaigning for clean skies and in the first few decades of the twentieth century the Smoke Abatement League specifically researched and campaigned to reduced smoke from the house chimney. In addition medical opinion was concerned about smoke from a public health perspective. Thus, for example, in 1899 a British Medical Association report found that the incidence of rickets was greatest in smoky areas and in 1908 the *British Medical Journal*, in a report on smoke pollution, argued that the domestic, rather than factory, chimney presented the greatest threat to public health.[64] But, as Otter points out, smoke as well as degrading public health also endangered the very basis of bourgeois liberal governmentality by undermining the visibility of civil conduct. The modern city was fashioned so that the 'respectable mastered their passions in public spaces conducive to the exercise of clear, controlled perception: wide streets, squares, and parks'.[65] The pollution of smoke threatened to return the city to an anti-bourgeois darkness in which conduct and behaviour could not be observed. Visual hygiene and the availability of clean air for the unhindered passage of sunlight was thus inseparable from the wider desires of the social hygiene movement. Sunlight not only brought health – it also revealed the open spaces of the city to the gaze of its controlling authorities: 'the moral, physical and biological fears they aroused [the slums] were inseparable from the spaces in which they lived. Liberal society, in short, needed to be built and maintained – wide streets, slum demolition, sewage and street lighting were all attempts to assemble spaces in where ruling freedom could be made possible and visible'.[66]

Be this as it may, the eventual passing of the Public Health (Smoke Abatement) Act of 1926 was a disappointment for the League and was described in *Sunlight* as a 'miserable triumph'. The act made no allowance for the control of smoke from domestic dwellings or for the empowerment of local authorities to control effectively either domestic or industrial smoke. Saleeby described it as a 'reactionary Act' and Chamberlain as having 'shown how little he had been acquainted with the facts... meanwhile, my advice to the unborn is HURRY NOT'.[67] It appeared that smoke was far more resistant to removal from the modern city than other forms of dirt. Otter contends that that the failure of smoke abatement in the early twentieth century 'suggests, something more than brute material resistance'.[68] In other words, the filthy city had itself been made into a symbolic bourgeois virtue with 'good business' and smoke being entwined. However, this somewhat underplays the 'brute materialities' which underpinned the smoke problem – if almost every home had solid fuel fireplaces as their main source of winter warmth, then change was no trivial matter. As Mosley points out, in the interwar years, smokeless fuels were more expensive, sometimes difficult obtain and often very difficult to use (e.g. coke is notoriously difficult to set alight). The alternatives to solid fuel (e.g. electricity or gas) were also often more expensive and incompatible with many existing domestic premises.[69]

5.5 Conclusions

Following the death of Saleeby in 1940, the Sunlight League fell into decline and ultimately ceased to exist during the Second World War. However, many of the developments outlined in this chapter facilitated the emergence of what we may term a nexus made up of sunlight, bodies and social environments. The parallel and indeed overlapping activities of the diverse groups discussed above share some commonalities. They all believed in: the moral or physical powers of nature to heal; the fear of the growth of 'unhealthy' urban conurbations; and the pejorative associations between 'darkness' and ill health (as contrasted with more positive views of sunshine and light). All of these, taken together with the developing sunlight therapies, helped to capture and stabilise the idea of the human body in sunlight, under a clear blue sky, as healthy. As might be expected, for this to happen required other related social and material changes. The promoting of the prophylactic exposure of the body to sunshine, by medical experts and the various organisations discussed above, partially legitimated the public exposure of the human body. In this regard the Sunlight League tirelessly campaigned in the interwar years for mixed bathing and an end to restrictive dress codes at seaside resorts and the new lidos.[70] Thus the growth of tourism and leisure pursuits from the late ninetieth to the early twentieth centuries, itself dependent on a series of social and technological changes, had a certain articulation with changing ideas about health and the tanned body. Once established, the idea of the suntanned body as healthy was to develop a strong resonance with the public, especially given the growth of mass tourism in the era after the Second World War – a resonance that is now being viewed in rather a different way by those with an interest in public health and preventative medicine.

Notes

1. B. Blacker and R. Clarke (1892). *Light as a Therapeutic Agent. The Practitioner*, XLVIII: 273–347.
2. S. Carter (2007). Rise *and Shine: Sunlight, Technology and Health* (Oxford: Berg).
3. F. Smith (1988). *The Retreat of Tuberculosis* (London: Croom Helm).
4. D. Kennedy (1990). 'The perils of the midday sun: climatic anxieties in the colonial tropics', in J. MacKenzie (ed.). *Imperialism and the Natural World* (Manchester: Manchester University Press), pp. 118–40.
5. D. Kennedy (2003). Diagnosing the Colonial Dilemma: Tropical Neurasthenia and the Alienated Briton. How Empire Mattered: Imperial Structures and Globalization in the Era of British Imperialism, unpublished conference paper, University of California, Berkeley.
6. C. Woodruff (n.d.). *The Effects of Tropical Light on White Men* (New York: Rebman Company).
7. R. Corson (1972). *Fashions in Makeup: from Ancient to Modern Times* (London: Peter Owen Ltd).
8. D. Mrozek (1987). 'The habit of victory: the American military and the cult of manliness', in J. Walvin (ed.). *Manliness and Morality: Middle-Class Masculinity in Britain and America 1800–1940* (Manchester: Manchester University Press), p. 227.
9. F. Mort (1987). *Dangerous Sexualities: Medico-Moral Politics in England since 1830* (London: Routledge & Kegan Paul).
10. J. Weeks (1981). *Sex, Politics and Society: the Regulation of Sexuality since 1800* (London: Longman).
11. A. Warren (1987). 'Popular manliness: Baden Powell, scouting and the development of manly character', in J. Walvin (ed.). *Manliness and Morality: Middle-Class Masculinity in Britain and America 1800–1940* (Manchester: Manchester University Press), pp. 199–220.

12. Ibid., p. 200.
13. Ibid., p. 212.
14. R. Baden-Powell (1922). *Rovering to Success: A Book of Life-Support for Young Men.* (London: Herbert Jenkins Ltd), p. 208.
15. C. Ward and C. Hardy (1986). *Goodnight Campers! The History of the British Holiday Camp* (London: Mansell Publishing Ltd), p. 3.
16. T. Holding (1908). *The Camper's Handbook* (London), p. 7.
17. C. Ward and C. Hardy (1984). *Arcadia for All: The Legacy of a Makeshift Landscape* (London: Mansell Publishing Ltd), p. 9.
18. Ibid.
19. Ibid., p. 11.
20. W. Morris (1974). *The Beauty of Life.* Abridged edn. (London: Brentham Press).
21. W. Morris (1962). *Art and the Beauty of the Earth* (Oxford: H.P. Smith).
22. E. Carpenter (1887). *England's Ideal and Other Papers on Social Subjects* (London: Swan Sonnenschein, Lowrey & Co), p. 75.
23. S. Sturdy and R. Cooter (1998). 'Science, scientific management, and the transformation of medicine in Britain c. 1870–1950', *History of Science*, xxxvi, 421–53; C. Lawrence (1998). 'Still incommunicable: clinical holists and medical knowledge in interwar Britain', in C. Lawrence and G. Weisz (eds). *Greater than the Parts: Holism in Biomedicine, 1920–1950* (Oxford: Oxford University Press).
24. Sturdy and Cooter, p. 451.
25. Interestingly many medical advocates of sunlight exposure, such as Henry Guavain and Auguste Rollier, argued for the need to strictly individualise medical treatment. See Lawrence, 'Still uncommunicable'.
26. Ibid.
27. Ibid., p. 16.
28. G. Rosen (1947). 'What is social medicine? A genetic analysis of the concept', *Bulletin of the History of Medicine*, 21, pp. 647–733.
29. G. Jones (1986). *Social Hygiene in Twentieth Century Britain* (London: Croom Helm).
30. R. Soloway (1990). *Demography and Degeneration: Eugenics and the Declining Birth-Rate in Twentieth Century Britain* (Chapel Hill: University of North Carolina Press).
31. Nethersole cited in Jones, *Social Hygiene*, p. 28
32. 'Health of the empire: "back to the light"', *The Times*, 26 May 1922, Sect. 7.
33. 'People's league of health: meeting at the House of Commons', *The Times*, 9 July 1924, Sect. 22.
34. Jones, *Social Hygiene*, p. 29.
35. 'Utra-violet rays and health: use of "vitaglass" at the zoo', *The Times*, 2 November 1926, Sect. 19.
36. The role of allotment gardening had a broader significance in working-class culture, not least economic. See R. McKibbin (1990). *The Ideologies of Class: Social Relations in Britain, 1880–1950* (OUP).
37. H. Belfrage (1926). quoted Jones, *Social Hygiene*, p. 29.
38. 'A new science', *The Times*, 22 May 1928, Sect. x.
39. H. Gauvain (1930). 'Sun treatment in England', *Journal of State Medicine*, pp. 38, 470.

40. J. Rosslyn-Earp (1929). 'Dosage in heliotherapy', *Journal of the American Medical Association*, pp. 92, 312.
41. 'The coming of the sunlight: influence on the nation's health, dispelling winter darkness', *The Times*, 22 May 1928, Sect. x.
42. 'Lens. saving the mothers', *New Statesman*, 8 December 1928, 32, p. 285.
43. G. Searle (2004). 'Caleb William Elijah Saleeby (1878–1940)', *Oxford Dictionary of National Biography* (Oxford: Oxford University Press).
44. 'Lens. the "Better dead" fallacy', *New Statesman*, 23 April 1923, 23, p. 250.
45. He was also a member of the Sociological Society and the Fabian Society.
46. 'Lens. a verdict for light'. *New Statesman*, 26 June 1920, 15, p. 330.
47. 'Lens. modern sun worship', *New Statesman*, 24 September 1921, 16, pp. 363–4.
48. 'Lens. modern sun worship II – its history', *New Statesman*, 8 October 1921; 17, pp. 10–1.
49. 'Lens. modern sun worship III – its high priest and his temple', *New Statesman*, 15 October 1921, 18, pp. 40–3.
50. Ibid., p. 40.
51. Ibid., p. 41.
52. 'Editorial foreword' (first edition). *Sunlight*, December 1924, 1 (1), pp. 5–8.
53. Ibid., p. 7.
54. Ibid., pp. 5–6.
55. W. L. Creese (1966). *The Search for Environment: The Garden City – Before and After* (New Haven: Yale University Press).
56. J. A. Burnett (1986). *Social History of Housing, 1815–1985* (London: Methuen), p. 219.
57. Ibid.
58. S. Mosley (2007). 'The home fires: heat, health, and atmospheric pollution in Britain, 1900–45', in M. Jackson (ed.). *Health and the Modern Home* (Abingdon: Routledge).
59. C. W. Saleeby (1928). 'From heliotherapy to helio-hygiene', *Sunlight*, December, I (7), p. 10.
60. Ibid., p. 9.
61. 'Editorial foreword: the sunlight view of the world', *Sunlight*, December 1926, I (3), pp. 3–8.
62. 'Smoke abatement: minister and sunlight league', *The Times*, 30 October 1926; Sect. 14.
63. E. Ashby and M. Anderson (1981). *The Politics of Clean Air* (Oxford: Oxford University Press).
64. Mosely, 'The Home Fires'.
65. C. Otter (2002). 'Making liberalism durable: vision and civility in the late Victorian city', *Social History*, 1 January 27 (1), 1.
66. Ibid., p. 3.
67. C. W. Saleeby (1927). 'A miserable triumph', *Sunlight*, July I (4), p. 14.
68. Otter, 'Making liberalism durable', p. 14.
69. Mosley, 'The home fires'.
70. This was partly aided by the introduction of elasticated fabrics that allowed the innovation of stretched, skintight swimsuits and greater exposure of the body. See F. Stafford and N. Yates, (1985). *Kentish Sources: IX. The Later Kentish Seaside (1840–1947)* (Gloucester: Alan Sutton Publishing Ltd).

6
Healthy Places and Healthy Regimens: British Spas 1918–50

Jane M. Adams

This chapter considers the use of climate and mineral waters for health maintenance and medical treatment in early twentieth-century Britain. These agents have been used since antiquity in a changing set of applications and therapies available at spas, and were increasingly exploited in Britain from the sixteenth to the mid-twentieth century.[1] However, their use declined dramatically after the end of the Second World War so that by the end of the twentieth century the specialised bathing facilities at British spas had all closed and spa treatments were no longer provided by the National Health Service (NHS).[2] Although a 'spa' industry continues to flourish, this is associated with leisure and beauty rather than health, and little emphasis is placed on the value of naturally occurring mineral-rich or thermal waters. This situation contrasts with approaches in some other European countries, notably Germany, where spa treatment continues to be available at specialist health resorts and active treatment under medical supervision is funded by the state healthcare system.[3] This chapter explores the background to current attitudes towards the use of climate and waters in Britain by examining ideas and practices in the interwar period and the attempts made to promote spa therapy to the public, the medical profession and the newly formed Ministry of Health (MH). The final section considers the factors that contributed to the decline in the use of spa treatments by the NHS.

By the interwar period spas were operating in an international market of specialist health resorts. Facilities at the major British spas were owned by local authorities or private companies, and financial viability depended on attracting paying patients. From the mid-nineteenth century access to spa treatment was expanded for the poorer classes through the mechanism of voluntary hospitals and other mutual and charitable organisations. The economic depression in the interwar period brought a recognition that future viability depended on securing support from the emerging state medical services. The activities of three organisations – the British Spa Federation (BSF) established in 1916, the International Society of Medical Hydrology (ISMH) formed in 1921 and the British Health Resorts Association (BHRA) set up in 1932 – were central to efforts made to secure this support. A key aim of the strategies adopted was the promotion of a specific identity for the major British spas that would differentiate them from seaside and pleasure resorts. Emphasis was given to their unique waters, natural environment, social amenities, treatment facilities and the role of specialist medical practitioners. In 1937 the BHRA handbook defined the cure as 'a set course of *spa treatment* in country surroundings, with a change of air and scene as well as diet and occupation', and noted that it depended on '*Waters, Climate, Methods*, to which must be added *Place*, the intangible *genius loci*, to which many ailing people are sensitive'.[4] Twelve British resorts were recognised by the BSF as meeting the criteria to be designated as spas; Bath, Bridge of Allan, Buxton, Cheltenham, Leamington, Droitwich, Harrogate, Llandrindod, Llanwrtyd, Trefriw, Strathpeffer and Woodhall.[5]

The discussion begins with an analysis of the concept of the health resort and the spa cure, constructs influenced by ideas drawn from a number of disciplines including medical geography, climatology and balneology that were debated and refined from the end of the eighteenth century. In the interwar period efforts were made to assert the importance of the medical specialist able to draw together clinical experience and scientific knowledge to harness the healing properties of water and climate. Consideration is also given to how the spa cure fitted within wider medical practice and the argument made that a number of cultural factors enhanced its appeal, notably a continuing interest in medical holism and an emphasis on regeneration of the nation. Early policy statements from the MH, established in 1919, which recognised the important contribution of environment and lifestyle in promoting good health endorsed the holistic approach that was an integral part of spa treatment. A national focus on preventive medicine and healthy lifestyles, together with recognition of the rights of all citizens to better health and improved access to healthcare, seemed to offer opportunities for spas to capitalise on their experience in these fields. The interwar period is also notable for new strategies used to promote spas using modern marketing techniques, including liaising with railway companies and collaborative initiatives including the 'Wintering in England'

and 'Come to England' movements, which targeted an international and national market, emphasising the potential health benefits of specific resorts.[6] The medical profession was recognised as being vital in channelling demand by recommending home resorts to their patients but marketing strategies had to strike a balance between effective publicity and professional protocols, which disapproved of advertising and self-promotion. The approach developed to circumvent these concerns was collaborative action co-ordinated through the BSF aimed at raising the status of medical hydrology with the medical profession and promoting knowledge about British spa resorts with good facilities for treatment. The BSF and the ISMH lobbied the MH aiming to secure support for expanding access to spa treatment to the less well-off, covered by the new National Health Insurance (NHI) scheme. The concluding section reviews attitudes to spa treatment in the early decades of the NHS and examines why, after an initial expansion in the numbers of patients treated at spas, confidence in the therapeutic potential of specialist resorts failed to maintain support.

6.1 Climate, waters, methods, place: the health resort and spa treatment

Interest in the relationship between climate and health had been rekindled from the late eighteenth century through the disciplines of medical geography and medical topography which studied the distribution of human diseases and the medical conditions of particular places. There were distinct national differences in the ways these ideas were conceptualised and applied in medical practice.[7] In Britain, influenced by trade and imperial ambitions, the utilitarian aim of keeping British soldiers and other Europeans healthy in tropical environments became a particular focus of attention. In India, for example, the coastal and temperate highland areas of the sub-continent were identified as being those most healthy for Europeans, but even these areas were deemed 'very inferior to corresponding climates in the temperate zone'.[8] Hermann Weber's remarks made at the end of the nineteenth century associating the British climate with vigour and health reflected prevailing imperialist assumptions about the superiority of European environments and populations.

A good climate is that in which all the organs and tissues of the body are kept evenly at work in alternation with rest. A climate with constant moderate variations in its principle factors is the best for the maintenance of health. It calls forth the energy of the different organs and functions, their power of adaptation and resistance, and keeps them in working condition. Such are the climates of England all year round, and they belong to the most health-giving in the world. They produce the finest animals, the finest trees, and the finest men and women, and are most conducive to health and longevity.[9]

Later writers concurred with Weber's view that the British climate was generally healthy, stimulating longevity and mental and physical vigour although a minority were more measured. T. D. Luke, for example, noted that the high humidity 'may have a sobering effect on the mentality' and contributed to the prevalence of rheumatic complaints.[10] The principle factors shaping climate were identified as latitude, altitude, proximity to the sea, geology and the prevailing winds.[11] The equable maritime climate of the British Isles was noticeably cooler in summer and warmer in winter than much of Europe, due to the influence of the Gulf Stream. The islands had high humidity, meaning there were many 'rainy days and rainy hours'.[12] Climate was believed to affect the whole metabolism and was described according to its relative stimulating effects on the body; tonic or bracing climates were contrasted with sedative or relaxing ones. A tonic climate was one with 'plenty of sunshine, little humidity, and a temperature which is comparatively cool but with a fair daily range'. In contrast a sedative climate was more humid with a smaller temperature range, higher rainfall and less sun. The therapeutic effects of a tonic resort were to stimulate 'the vigour of the body, thus enabling it to combat or overcome the onslaught of disease', while sedative resorts were suited to more delicate individuals, including the very young and old.[13]

Within the British Isles there were accepted regional variations in climate, justified by extensive comparative statistics detailing meteorological information including temperature, winds and hours of sunshine. These statistics showed the east and south-eastern seaside resorts to be drier and colder, and therefore more bracing, than those in the south-west. However, it was a general feature of the whole country that there was 'almost regular alteration of tonic and sedative types of weather, or what might be described as a happy blend of bracing and relaxing elements' although the 'local complex of climate and landscape' was also important in determining tonicity, especially the degree of exposure to the elements and the availability of shelter from wind and rain.[14] Britain was characterised by many seaside resorts and relatively few inland ones in contrast to Europe, where the main resorts were the inland spas and mountain resorts and where the seaside remained relatively underdeveloped until the interwar period.[15]

In the second half of the nineteenth century, the theoretical rationale for the use of natural waters was also transformed. Prior to this, natural mineral-rich or thermal waters were used in simple ways – drinking or bathing by immersing the whole body – with limited use of water jets or friction massage. The active healing agents were not fully understood, with significance given to their distinctive chemical and thermal attributes rather than to the generic properties of water.[16] Spas developed reputations for the treatment of particular diseases; for example the effect of Cheltenham waters, taken internally by drinking, was to encourage bowel action and they were widely used for liver complaints; whereas Buxton water, used for both bathing and drinking, was deemed useful for rheumatism. From the late 1820s onwards hydropathy, first associated with the cold water cure of Vincent Preissnitz, introduced a different approach, using pure water in a wide variety of applications including local baths and showers.[17] Although initially promoted as a separate therapeutic approach, by the 1880s, due in large part to research carried out at European spas and by Wilhelm Winternitz at Vienna, orthodox practitioners specialising in water treatments began to draw on both spa medicine and hydropathy to harness the potential benefits of waters and baths.[18] Although uncertainties remained in identifying the specific therapeutic action of mineral content, dissolved gasses or temperature, clinical experience was asserted as sufficient evidence for their effective use in treating a variety of conditions including gout, constipation, nerves and neurasthenia.[19]

Spa practitioners were first considered as experts in the use of specific local waters, their reputation based on empirical knowledge developed through extensive clinical experience. However, by the 1880s they were also presenting themselves as specialists in the medical use of all waters. The Society of Balneology and Climatology, established in 1885, provided a forum for discussion and presentation of research through the *Journal of Climatology and Balneology* and in 1906 was recognised as a separate section within the Royal Society of Medicine. Robert Fortescue Fox, the most prominent British figure in twentieth-century spa medicine, promoted the use of the term, 'medical hydrology', which he defined as 'the science of waters, vapours, and mineral deposits in connection with waters, as used in medicine, both by internal administration and in the form of baths and applications'.[20] Fox was instrumental in setting up the ISMH in 1921 becoming the first editor of its journal, the *Archives of Medical Hydrology*. The society's aims included coordination of international research into the therapeutic effects of mineral waters, development of a common scheme of classification and promotion of their use at specialist resorts at which healing was emphasised over pleasure. A definition put forward for discussion by members in 1939 highlighted the importance of medical expertise in this model.

> A spa is a place where there occur mineral waters or natural deposits of medicinal value, where the local administration has provided suitable facilities for making use of these, and where the treatments are given under medical direction.[21]

The emphasis given to the combination of natural waters, specialist facilities and clinical experience was intended to clearly differentiate spas from other resorts, particularly the many seaside settlements which had emerged as centres for leisure and holidays in Britain during the nineteenth century. As Fox noted:

> A place for the recovery of health owes its virtue to many things- to the good gifts of nature, to science with its wise application of natural forces, to medical personality, to tradition and many more.[22]

The factors involved in spa treatment were wider than the techniques of medical hydrology, extensive though these were. In a lecture on the principles of spa treatment given in 1929 Wilfred Edgecombe, one of the honorary physicians at Harrogate, drew attention to the close association between mind and body, noting the need to consider what he termed 'both physical and psychical' aspects in developing treatment. In addition to waters, baths and climate the other physical agents included massage, exercise, mechano-therapeutics and electrical procedures as well as diet and general regimen. The 'psychical' factors, deemed capable of improving the mental state of the patient, were listed as change of scene or environment, absence from business worries and recreation and general amenities.[23] Given the multitude of factors involved it is not surprising that the specific therapeutic action of individual factors remained unclear, but despite these uncertainties the efficacy of spa treatment remained credible. As Edgecombe argued, cumulative clinical experience supported the view that it 'has a real utility in the management, both preventive and curative, of chronic disorders'.[24]

While emphasising that 'the waters must of course, take the first place as a means of cure', the BSF's annual publication recognised the importance of climate and auxiliary treatments as well as social amenities. The British spas offered a considerable range of settings – Bath, Cheltenham and Leamington were the largest, all important residential towns by the start of the twentieth century, while at the other end of the scale Woodhall Spa was little more than a village. Brief remarks on their comparative climates described all the British spas as more tonic than the European spas with Bath, Cheltenham, Droitwich and Leamington the most sedative, while at the other end of the scale Harrogate and Buxton were denoted 'frankly tonic'.[25] Identification of the special healing properties of waters remained elusive and by the turn of the twentieth century attention had turned away from chemical composition towards radioactivity as a potential explanation for their therapeutic effectiveness; this property was widely featured in advertising. The brief guide to each spa included an analysis of the waters, their principle action on the body and a note on the main diseases treated, which included nervous fatigue, toxaemias, circulatory disorders, rheumatism, convalescence and anaemia.[26] In addition to providing details of the bathing establishments and pump rooms, descriptions of social amenities and amusements were also included, as facilities for relaxation and recreation were fundamental in shaping the social environment of a spa and were deemed capable of a profound influence on physical and mental health.

Fresh air was also an important part of the regime, and parks and gardens with ample seating were provided everywhere. Sports facilities also expanded from the end of the nineteenth century and by the early twentieth century commonly included archery, riding, walking, croquet, lawn tennis, boating, swimming, ice- and roller-skating, angling, golf and cycling.[27] Several resorts held sports tournaments, such as the North of England Croquet Championship in Buxton, while Leamington emphasised its 'three packs of foxhounds and one of beagles' within easy reach.[28] During the interwar years investment in facilities for sports and open-air recreation continued: facilities at Woodhall were transformed with the development of a lido and park in the 1930s funded by Lady Weigall the major benefactor of this small resort. Lidos were also constructed in Cheltenham and Droitwich in 1935, providing access to water for swimming and leisure purposes to complement the treatment facilities at the medical baths. Music was provided at most spas for part of the day during the season, while the larger centres at Bath and Buxton provided opera and a resort orchestra. Spas also invested in the latest entertainment facilities, so that even a small resort such as Woodhall had a cinema by the 1920s.[29] The mix of social amenities and leisure facilities at spas were similar to those at many seaside and inland resorts, what differentiated them was specialist medical bathing institutions.

Although the rise of biomedicine, associated with the success of technology and the laboratory, remains the dominant narrative of medical history from the mid-nineteenth century, a continuing interest in medical holism is also apparent. The persistence of these ideas was partly due to support from influential members of the London elite whose focus was on clinical practice rather than research and who endorsed the value of 'collective clinical experience'.[30] Despite the enormous growth in medical knowledge, relatively few effective therapies had been developed and although there had been 'a few striking successes, such as salvarsan, vaccine therapy, and insulin', many diseases remained difficult to treat. These practitioners valued the 'vis mediatrix naturae', or healing power of nature, which regarded the body as having a natural tendency to recover from disease and considered it a key objective of medical treatment to encourage this. The 'vis mediatrix naturae' had an ancient origin and had also been closely associated with hydropathy and the other alternative medical systems of the mid-nineteenth century. Other forms of the 'holistic turn' in the interwar period included recognition of the profound effects of physical and social environment on health, the perception of sickness as a general disorder of the body and a willingness to treat the mind and the body together and 'incorporate emotions and the psyche into the study and cure of individuals'.[31] Several members of this elite group of clinicians, including Lord Horder and Lord Dawson, were active supporters of the British spas.

From the eighteenth century onwards spa treatment, with its emphasis on all aspects of regimen, had been used to combat the supposed deleterious effects of civilisation on health, perceived to affect the nerves in particular. Associated with politeness, affluence and excess in the eighteenth century, by the 1870s nervous diseases were linked with the stresses of modern life in industrial society.[32] This was articulated most clearly by George Miller Beard who popularised a new term, neurasthenia, which he argued arose from the unique stresses of modern day American life. Although neurasthenia did not achieve the same popularity as a diagnosis in Britain as in America, it gained some currency from the 1880s onwards partly influenced by the experience of shell shock in the First World War.[33] The treatment of wounded and convalescent servicemen with hydrological and physical methods undoubtedly raised the profile of these therapies with both the public and the medical profession and this, together with restrictions on foreign travel and heightened feelings of nationalism, encouraged patients and visitors to use the home resorts. An article in *The Times* in 1915 reflected a growing consensus that the British spas had both the natural resources and the expertise and facilities to match those available elsewhere in Europe, promoting the concept that they were a 'national asset'.[34] The wider cultural project of national regeneration and renewal prevalent in the interwar period offered further opportunities to capitalise on this support.

Belief in the prophylactic benefits of a healthy lifestyle integral to spa regimes was in tune with the founding principles of the MH, established in 1919.[35] In *An Outline of the Practice of Preventive Medicine*, Sir George Newman, the first Chief Medical Officer, emphasised the influence of physical and social environment on health and the need for measures to address the burden of illness across all social classes through preventive and curative medicine. One example of this was state support for the expansion of sanatoria treatment for tuberculosis. For Newman, health was also linked with morality and citizenship as a strong nation was made up of healthy individuals. Ignorance and poor self-control contributed to poor health among the masses so that in addition to providing better access to GP and hospital services, measures were put in place to encourage healthy lifestyles on the basis that effective preventive medicine would improve the health of the population and reduce disease.[36] The Central Council for Health Education, formed in 1927, promoted the idea that health was not just an individual benefit but an obligation of citizenship as it was the responsibility of all 'to make a very real contribution to the National well-being'.[37]

6.2   The image of the health resort: promoting British spas

The wartime conditions that had led people to frequent the home resorts proved to offer only temporary respite from foreign competition; as travel restrictions were relaxed and European resorts re-opened, spa managers took a range of measures to attract patients and visitors in deteriorating economic conditions. In 1921 the Health Resorts and Watering Places Act, the result of several decades of lobbying, allowed towns to use the profits from municipal enterprises up to the equivalent of a penny rate to fund advertising in newspapers, handbooks, leaflets and railway placards. This inaugurated a period of unprecedented marketing and promotional activity.[38] Further legislation in the 1930s gave increased powers to local authorities but, unlike its counterparts in France and Belgium, the British state did not take a lead in promoting tourism. Instead developments were pushed through by a number of campaigns and associations drawing together interested parties. John Beckerson describes the 'Come to Britain' Movement set up in 1926 as using 'a well tried model of voluntary coalition of trade interests, led by the great and the good. Shipping lines, railways, London stores and the Association of Health and Pleasure Resorts were all represented'.[39] Although many of these collaborative ventures advertised Britain as a tourist destination, particularly in the context of an international market, this was also a period when images for particular towns were created through the use of slogans and posters. Cheltenham sought to exploit associations with prestigious European spas with publicity material emphasising its 'continental' character and coined the description the 'English Carlsbad'. In contrast, Buxton highlighted its natural surroundings and promoted itself as the 'Mountain Spa'.[40]

The scope of promotional activity in this period is illustrated by measures put in place at Leamington co-ordinated by W. J. Leist, who in 1924 was appointed manager of the Pump Rooms, which were owned by the local authority. Of the annual budget of £400, £250 was allocated to initiatives with Great Western Railway, £45 to advertising in various publications, £77 paid to the BSF and £28 to the ISMH.[41] Collaborative advertising with railway companies was an important innovation, funding illustrated tourist brochures and colour posters distributed by the companies and displayed across the railway network.[42] Great Western Railway was renowned for its 'impressive range of handbooks guides'. A 1923 publication promoted the various resorts on the network together with details of tourist and weekend railway tickets.[43] The entry for 'Leafy Leamington' emphasised the genteel urban pleasures of the town, including the excellent shops, the beauties of the Jephson Gardens and range of all year entertainments, in addition to the up-to-date treatment facilities.[44] Posters for the town featured a middle-class clientele enjoying the waters, gardens and sports facilities in the Pump Rooms and adjacent Jephson Gardens. The tenor of this publicity, aimed at the general public, continued to promote the wider attractions of the spas, giving emphasis to both social amenities and health facilities.

Treatment facilities at Leamington were substantially upgraded in the 1920s and the opening of the refurbished Pump Rooms was used to publicise both the spa centre and the town. Plans for redevelopment had been informed by visits to Bath and Cheltenham to assess facilities there; among the improvements made were the introduction of several new treatments including paraffin wax, Berthollet Steam apparatus,

Mountain Sun treatments and specialist masseurs.[45] After a variety of delays compounded by the General Strike, the baths were opened in October 1926. The invited audience included medical staff from Queen's Hospital in Birmingham and the local Warneford Hospital; coverage of the event featured in the national and medical press including the *Lady, Queen, Lancet* and the *BMJ*.[46] Sir Kingsley Wood, Parliamentary Secretary to the MH, formally opened the facilities, commenting in his address that 'he thought the time had come for British spa treatment to be made available for all classes of the community' and that the MH would be giving 'careful and sympathetic consideration' to a scheme under development by the BSF and the Friendly Societies.[47] These proposals to expand access to spa treatment to the lower paid are discussed further below.

The Pump Rooms also issued its own publicity directed at the public and the medical profession, producing two separate booklets tailored for each audience. Ten thousand copies of a general booklet, financed through the sale of advertising space, was produced and distributed through a sophisticated network that included Great Western Railway, Cunard, Thomas Cook and several other large shipping lines and travel bureaux. A copy was also sent to every doctor in Warwickshire.[48] A smaller run of 2000 copies of a specialist medical booklet was printed and distributed to doctors in the surrounding towns, as well as to a selection of practitioners in London and the provinces.[49] This was a more sober affair, with no advertisements and more extensive information on the waters, treatment facilities and process for referring patients. Regular advertisements for the spa were also placed in newspapers and publications carefully selected to target the wealthy middle-class market that was the mainstay of the spa's prosperity. These included *The Times, Observer, Daily Telegraph, Morning Post, Daily Mail, Daily Chronicle, Observer, Sunday Times, Church Times, Queen, Lady* and the *Gentlewoman*.[50]

Publicity policy at spas was constrained by the conventions of professional behaviour, which deprecated advertising of the services of individual practitioners and the promotion of the therapeutic effects of particular waters.[51] An important vehicle for reaching this audience was the annual handbook produced by the BSF, a publication which promoted the use of spa therapy in general as well as providing practical details for the referral of patients, and which was widely distributed among the medical profession. The BSF also represented the interests of spas within organisations with a wider remit, such as the BHRA which aimed to use 'all legitimate ways in which publicity may be given to British Health Resorts' and to develop the understanding of the medical profession and the laity of their potential therapeutic uses.[52] The directors of the BHRA included several influential members of the medical elite, including Lord Horder, Sir Humphrey Rolleston and W. G Willoughby, President of the Society of Medical Officers of Health and later President of the BMA.[53] Re-iterating a long-held belief in the detrimental effects of modernity on health, the BMA's literature suggested that the sedentary life of town dwellers, compounded by the 'nervous stresses inseparable from civilisation' could result in the 'fatigue-factor' that contributed to increased pathology.[54] A well directed course of treatment at a health resort, it was argued, would improve overall health and mitigate these deleterious effects. The BMA funded an annual guide, *British Health Resorts,* compiled by Fortescue Fox, which provided brief details of member resorts and organised the contribution of promotional articles to the lay and medical press. It held regular regional conferences on topical issues of both general and medical interest. The 1937 conference included sessions on 'The Spa as a National Asset in the Maintenance of Physical Fitness', opened by Lord Horder and a technical forum on 'Hydrological methods in the treatment of injuries and diseases of the joints' opened by Dr F. C. Thomson, physician at the Royal United Hospital Bath and a former chairman of the BMA. The BMA was also active in promoting the value of visiting resorts outside of the summer season and took an interest in improving standards by setting up a system of visits and inspections in 1935. At the heart of these promotional activities lay the message that the health resort had something to offer patients seeking treatment for specific illnesses or needing a period of respite from the strains of everyday life. This was neatly summed up in the dedication in the front of T. D. Luke's *Spas and Health Resorts of the British Isles*, published in 1919, which was addressed 'To those who *enjoy* poor health – to those who are sick – to those who are well and wish to keep so'.[55]

6.3   Spas and the state

Articles in the early issues of the *Archives of Medical Hydrology* demonstrated there were considerable differences in government support for hydrology across Europe.[56] A particular weakness in Britain was identified as the lack of a national body analogous to the French Commission des Eaux Minerales, which assessed all mineral waters within a national classification system and validated their therapeutic effects. Other comparative shortcomings included few facilities for research and limited emphasis in the medical curriculum. The ISMH and the BSF lobbied the MH for government support for these issues throughout the interwar years but to no effect.[57] Despite informal support given to a number of organisations promoting medical hydrology and spas discussed above, the MH endorsed the British tradition of reliance on local arrangements and initiatives.[58] This approach was consistent with other areas of the MH's work, the limited development of health centres in the interwar period being a case in point. Despite growing consensus that the state should take an active role in facilitating healthy lifestyles and encouraging physical fitness, the appropriate roles for the central state, local authorities and voluntary and private sectors remained unclear.[59] In relation to medical education, a group of spa practitioners formed the Committee for the Study of Medical Hydrology in Great Britain to promote teaching in medical schools and a panel of lecturers was formed to provide lectures to medical students and medical societies.

The BSF also lobbied for spa treatment to be funded by the newly introduced NHI scheme.[60] This campaign centred on treatment for rheumatism, which had been identified as a significant health priority, accounting for 'nearly one-sixth of the industrial invalidity' with over 3,000,000 weeks of working time lost each year from the insured population at a cost of almost £2,000,000 in sickness benefit.[61] Funding was requested to supplement existing access routes for the poor and low paid workers through the voluntary mineral water hospitals and a range of other institutions funded by friendly and mutual associations.[62] For example the Birmingham Hospitals' Saturday Fund ran Highfields Hospital in Droitwich and St Ann's Orchard in Malvern for its members. In 1925 the United Patriot's Approved Society, supported by representatives from the BSF put forward a detailed case for spa treatment for rheumatism to be added to the list of statutory treatments to the Royal Commission on National Health Insurance. Their submission claimed that 17.9% of the organisation's expenditure between 1921 and 1923 was linked to rheumatic disease.[63] The scheme proposed using spare capacity at spas to provide outpatient treatment for 1150 patients in summer and 1550 in winter, with accommodation in approved lodgings or purpose-built hostels. Total cost per patient was estimated at £7 19 shillings for a three-week course including baths, massage, electrical treatments, and board and lodging. Although these direct costs would be offset by savings on sickness benefit as patients returned to work more quickly, it was recognised that some additional Treasury grant or use of surpluses in the valuation pool would be needed to fund the proposal.[64] When the Commission turned down the application for a statutory benefit the BSF and the Approved Societies began to lobby for spa treatment to be added to the list of additional benefits, able to be provided by individual societies if their funds allowed.

The alternative proposal to providing treatment at spas was to establish 'town clinics' for outpatient treatment, similar to a model developed in Berlin, where hydrological and radiant light and heat treatments were provided at the end of the working day.[65] At a meeting held at the MH in March 1927, intended to shape investment decisions, the British Council on Rheumatism spoke in favour of town clinics with well-equipped hydro-therapeutic departments serving a local popula- tion, while representatives of the Spa Practitioners Group and the BSF supported developing facilities at established spas.[66] The concept of town clinics using any clean water, posed a fundamental threat to the justification for spas which, as has been shown, based their claims to be specialist centres not only on methods of application but also on their natural waters and the benefits of a stay at a specialist health resort. Town clinics offered a fundamentally different model of care, providing treatment using a variety of applications on an outpatient basis incorporated into the daily routine of the worker. The discus- sion also highlighted social and cultural issues; Neville Chamberlain, for the MH, noted that one objection to the 1925 proposal was of inequity, commenting that 'a statutory benefit could not be defended when only open to insured persons during a slack season'. The support of the Spa Practitioners Group for special hostels was influenced by a concern to exercise control over patients' behaviour as 'if they were in private boarding houses they would probably spend their evenings in public houses'.[67] The 1925 proposal had alluded to this issue but concluded that 'Approved Society members are responsible persons who will generally be anxious to get better, and they will do all they can to help forward the treatment'.[68] In February 1929, in readiness for the next quinquennial review of NHI, the BSF proposed a scheme at a fixed price of £9 6 shillings for a three-week stay for an estimated 2000 patients to be funded equally by the Approved Society and the patient. Accommodation was to be in special hostels and appropriate behaviour enforced by the threat of termination of treatment for non-compliance with the prescribed regime.[69] Despite this sustained lobbying over several years, the financial pressures on NHI resources meant the MH continued to oppose the inclusion of spa treatment as an additional benefit on the basis it 'might have the effect of reducing the amount of money allocated by Societies for expenditure on the present benefits, or of [approving] costly benefits which only a few wealthy Societies would be able to adopt, and which would, therefore, be available to only a small proportion of insured persons'.[70]

It is significant that the case articulated against spa treatment was not that it was ineffective but rather that it was not affordable. Class considerations were also apparent, the use of spare capacity in the off-season offended democratic principles, while concerns about inappropriate behaviour reflected the concern of practitioners not to alienate middle-class paying customers. Despite the failure to approve funding for spa treatment from the NHI scheme, there are indications that the MH considered that hydrological and physical methods offered potential for mitigating the burden of rheumatic disease. In 1929 J. Alison Glover, representing the MH, had expressed support for its use in his opening address to the annual meeting of the ISMH in London, the theme of which was 'The relation between the state, the health resort and national health insurance' in various countries.[71] On the one hand spas were seen as places offering expertise in treating rheumatism but the cost of the spa cure, which rested on attendance at a specialist health resort for three weeks, was more than the NHI scheme could bear. The alternative of local clinics, close to where people worked and lived, was a better fit with the model of local health centres close to people's homes. Some argued that money was better spent on prevention than expensive treatments; although the causes of rheumatic disease remained obscure one current theory suggested local infection, in particular dental sepsis, could be an important factor so that NHI funds were better spent on dental benefit as a preventive strategy.

Despite the lack of success in achieving support from the NHI scheme, spa treatment was supported by some individual mutual societies. In 1927 the BSF negotiated a scheme with the Post Office for officers earning under £150 a year in the provinces and £160 a year in London to access spa treatment for an inclusive weekly charge of 18 shillings. Costs of board and accommodation were on top of this, payable by the officer. In 1928 a similar scheme was negotiated for staff at the Ministry of Labour.[72] Arrangements to attract more working and middle-class patients also continued at local level. In Buxton, for example a new clinic aimed at the middle class was opened in 1935, providing accommodation and treatment for between £4 4 shillings and £6 6 shillings a week. The official opening took place during one of the BHRA conferences, with the opening speech given by Lord Horder.[73] The clinic proved a success and was extended in 1938, by which time there were plans for a combined development with the Devonshire Royal Hospital and the Empire Rheumatism Council.

6.4   Spa treatment under the NHS

This discussion has shown that despite the lack of support for British spas from the MH a holistic model of spa treatment continued to attract support from private patients, individual clinicians and a variety of organisations involved in funding healthcare throughout the interwar period. Although these services were not specifically included in planning for the NHS, treatment at several British spas did become available after 1948, accessed principally through the voluntary hospitals which were brought within the state sector. At several spas – including Harrogate, Buxton, Leamington, Droitwich and Woodhall – the main hydrological treatment facilities were in separate bathing establishments owned either by the local authority or private companies. Access to these facilities for NHS patients was secured by local agreements between the hospital boards and the relevant authorities. For example at Harrogate, NHS patients were accommodated at the Royal Bath Hospital but took treatments at the Royal Baths. Treatments provided under the NHS increased after the Leeds Hospital Board invested in additional accommodation in the town at the White Hart Hotel. In contrast the number of private patients declined from 62,000 in 1948–9 to 13,000 in 1962–3.[74]

A report prepared by the Spa Practitioners Group Committee of the BMA under the chairmanship of Lord Horder in 1951 highlighted several trends 'hastened by the inclusion in the National Health Service', notably less emphasis on 'the presence of natural medicinal waters or their specific therapeutic value' and more on a wider group of treatments applied under the supervision of physiotherapists.[75] Although this included hydrotherapy, by the post-war period this term implied active exercise in large therapeutic pools with no importance attributed to mineral content. The cost of funding treatment in old bathing establishments, several of which dated back to the early nineteenth century and had been built with luxurious facilities to attract elite patients, also contributed to the later withdrawal of the NHS from contracts with spa authorities. Purpose-built hydrotherapy and physiotherapy departments situated closer to other clinical facilities eventually replaced the old spa buildings, although this not happen quickly, for example physiotherapy services were provided from the Royal Pump Rooms at Leamington until 1997. However, in contrast to the definition of a spa noted earlier, treatment regimes after 1948 can be described as emphasising methods rather than climate, water or place. When the Royal Devonshire Hospital at Buxton finally closed in 2000, one of the reasons given was the need to move services closer to where people lived; it appeared that the concept of a specialist resort harnessing unique natural assets did not fit easily with new aspirations such as the provision of local services and equal access.[76]

Notes

This article has evolved from a Wellcome Trust funded project 'Healing cultures, medicine and the therapeutic uses of water in the English Midlands, 1840–1948' (Grant No. 077552/Z/05/Z/AW/HH). I would like to thank the Trust for supporting this work. The chapter has benefited from comments from Hilary Marland and the editors; any shortcomings remain, of course, my own.

1. Useful surveys are R. Porter (ed.) (1990). *The Medical History of Waters and Spas, Medical History*, Supp. No. 10 (London: Wellcome Institute for the History of Medicine); P. Hembry (1990). *The English Spa 1560–1815* (London: Athlone); P. Hembry (1997). *British Spas from 1815 to the Present*, edited and completed by L.W. Cowie and E. E. Cowie (London: Athlone).
2. W. A. R. Thomson (1978). *Spas that Heal* (London: Adams and Charles Black).
3. G. Weisz (1990). 'Water cures and science: the French academy of medicine and mineral waters in the nineteenth century', *Bulletin of the History of Medicine*, 64, pp. 393–416, 393–5; T. W. Maretzki (1989). 'Cultural variation in biomedicine: the *Kur* in West Germany', *Medical Anthropology Quarterly*, 3.1, pp. 22–35.
4. R. Fortescue Fox (ed.) (1937). *British Health Resorts: Spa, Seaside, Inland*, 5th edn (London: J. & A. Churchill), p. 15.
5. 'The British spas: a medical Appreciation', *British Medical Journal (BMJ)*, 11 June 1932, p. 1095.
6. J. Beckerson (2002). 'Marketing British tourism: government approaches to the stimulation of a service sector, 1880–1950', in H. Berghoff, B. Korte, R. Schneider and C. Harvie (eds). *The Making of Modern Tourism: The Cultural History of the British Experience* (Basingstoke: Palgrave Macmillan), pp. 133–57, p. 141.
7. N. A. Rupke (ed.) (2000). *Medical Geography in Historical Perspective, Medical History*, Supp. No. 20 (London: Wellcome Institute for the History of Medicine) provides a useful introduction.
8. J. Murray (1844). 'Practical observations on the nature and effects of the hill climates of India', *Transactions of the Medical and Physical Society of Bombay*, 7, pp. 79–154, p. 6, quoted in M. Harrison (2000). 'Differences of degree: representations of India in British medical topography, c. 1820–1870', in Rupke (ed.). *Medical Geography*, pp. 51–69 on p. 62.
9. H. Weber and F. P. Weber (1907). *Climatotherapy and Balneotherapy: The Climates and Mineral Water Health Resorts (Spas) of Europe and North Africa* 3rd edn (London: Smith Elder), p. 235. This paragraph was reprinted in several works, the earliest in an address to the British Balneological and Climatological Society in April 1899 reprinted in the *Lancet* 20 May 1899.
10. T. D. Luke (1919). *Spas and Health Resorts of the British Isles* (London: A. & C. Black), p. 49.

11. Weber and Weber (1907). *Climatotherapy and Balneotherapy*, p. 16.
12. Ibid., p. 233.
13. E. Hawkins (1923). *Medical Climatology of England and Wales* (London: H. K. Lewis), pp. 257–9.
14. L. C. W. Bonacina (1937). 'Climate, health and the British resorts', in Fortescue Fox (ed.) *British Health Resorts*, p. 12.
15. J. K. Walton (1983). *The English Seaside Resort: A Social History 1750–1914* (Leicester: Leicester University Press) and J. K. Walton (2000). *The British Seaside: Holidays and Resorts in the Twentieth Century* (Manchester: Manchester University Press).
16. C. Hamlin (1990). 'Chemistry, medicine and the legitimization of English spas, 1740–1840', in Porter (ed.). *Medical History of Waters and Spas*, pp. 67–81.
17. R. T. Claridge (1842). *Hydropathy or the Water Cure as Practised by Vincent Preissnitz* (London: James Madden).
18. J. M. Adams (forthcoming). *Healing with Water: Spas and the Water Cure in Modern England*, especially Chapter 1, 'The Theory and Practice of the Water Cure'.
19. See, for example, R. Porter and G. S. Rousseau (1998). *Gout: the Patrician Malady* (New York: Yale University Press); J. C. Whorton (2000). *Inner Hygiene: Constipation and the Pursuit of Health in Modern Society* (Oxford: Oxford University Press) and J. Oppenheim (1991). *'Shattered Nerves': Doctors, Patients and Depression in Victorian England* (Oxford: Oxford University Press).
20. R. Fortescue Fox (1913). *The Principles and Practice of Medical Hydrology: Being the Science of Treatment by Waters and Baths* (London: London University Press), p. 259.
21. Anon. (1939). 'What is a Spa?' *Archives Medical Hydrology*, XVII. 3, 124.
22. R. Fortescue Fox (1937). *British Health Resorts*, p. 4.
23. W. Edgecombe (1929). 'A lecture on the principles of spa treatment', *BMJ*, 1 June, pp. 981–83.
24. Ibid., p. 983.
25. R. Fortescue Fox (n.d. but after 1923). 'Introduction', in *The Spas of Britain: The Official Handbook of the British Spa Federation* Bath: Pitman Press), pp. v–xiv.
26. Ibid., p. xiii.
27. K. E. McCrone (1988). *Sport and the Physical Emancipation of English Women, 1870–1914* (London: Routledge), p. 249.
28. *Spas of Britain* (n.d. but after 1923), pp. 39, 99.
29. Ibid., p.147.
30. C. Lawrence and G. Weisz (1998). 'Medical holism: the context', in C. Lawrence and G. Weisz (eds). *Greater than the Parts: Holism in Biomedicine, 1920–1950* (Oxford: Oxford University Press), pp. 1–22.
31. Ibid., p. 2.
32. R. Porter (2001). 'Nervousness, eighteenth- and nineteenth-century style: from luxury to labour', in M. Gijswijt-Hofstra and R. Porter (eds). *Cultures of Neurasthenia: from Beard to the First World War* (Amsterdam and New York: Rodopi), pp. 31–48, p. 42.
33. M. Thomson (2001). 'Neurasthenia in Britain: an overview', in Gijswijt-Hofstra and Porter (eds). *Cultures of Neurasthenia*, pp. 77–95, p. 78.
34. 'The English spas – Buxton', *The Times*, 6 March 1915, p. 11.

35. S. Sturdy (1998). 'Hippocrates and state medicine: George Newman outlines the founding policy of the ministry of health', in Lawrence and Weisz, *Greater than the Parts*, pp. 112–34 on pp. 112–3.
36. G. Newman (1919). *An Outline of the Practice of Preventive Medicine: A Memorandum Addressed to the Minister of Health* (London: HMSO).
37. National Fitness Council (1939). pp. 4–5, 24, quoted in I. Zweiniger-Bargielowska (2007). 'Raising a nation of "good animals": the New Health Society and health education campaigns in interwar Britain', *Social History of Medicine*, 20 (1), 73–89 on p. 86.
38. The National Archives (TNA): Public Record Office (PRO), HLG 52/114 and 52/115.
39. Beckerson (2002). 'Marketing British tourism', p. 141.
40. Luke (1919). *Spas and Health Resorts*, Advertisements, p.1 after p. 318.
41. M 4348.1995: 1916–1927, Royal Leamington Spa, Pump Room and Baths, Manager's Report Book, June 4th 1924.
42. S. V. Ward (1998). *Selling Places: The Marketing and Promotion of Towns and Cities 1850–2000* (New York: Routledge), p. 35.
43. B. Cole and R. Durack (1992). *Railway Posters from the Collection of the National Railway Museum, York* (London: Lawrence King), p. 9.
44. Anon. (1923). *Spas and Inland Resorts: Health Giving Centres in the Midlands, the West of England and Central Wales* (London: Great Western Railway). This included Bath, Cheltenham, Church Stretton, Droitwich, Leamington Spa, Malvern, Torquay, Weymouth and the Welsh spas.
45. Royal Pump Rooms, M 4151. 1994, Medical Advisory Pump Rooms Committee: Minutes 1919–1959, 1 May 1924, 10 September 1924, 27 March 1927.
46. Ibid., 22 July 1926.
47. 'Reconstruction of the Leamington Pump Room', *BMJ*, 16 October 1926, p. 709.
48. Royal Pump Rooms, M 4151. 1994, Medical Advisory Pump Rooms Committee: Minutes 1919–1959, 23 April 1925.
49. Ibid., 26 June 1925.
50. Ibid., 23 April 1925.
51. Anon. (1935). 'Medical ethics for spa practitioners', *Archives Medical Hydrology*, XIII, 3, 51.
52. Annual Report of the British Health Resorts Association 1937/8, p. 2. Twenty-four of the 40 directors were medical practitioners.
53. TNA: PRO. BT31/35556/267466. Annual Return of the British Health Resorts Association, 19 October 1932.
54. 'British Health Resorts Association: first spa conference', *BMJ*, 28 May 1932, p. 1001.
55. Luke (1919). *Spas and Health Resorts*, dedication.
56. These ran under the heading 'The present state of medical hydrology in the countries represented' throughout 1922–4.
57. TNA: PRO. MH/58/159, 'Statement for the Ministry of Health, from the Joint Committee (of the International Society of Medical Hydrology and the Section of Balneology of the Royal Society of Medicine) to PROMOTE THE TEACHING OF MEDICAL HYDROLOGY IN ENGLAND', November 1923.

58. TNA: PRO. MH 55/1044, Minute Sheet to CMO from J. A. Glover, dated 6 February 1934.
59. A. Beach (2000). 'Potential for participation: health centres and the idea of citizenship c. 1920–1940', in C. Lawrence and A.-K. Meyer (eds). *Regenerating England: Science, Medicine and Culture in Inter-War Britain* (Amsterdam and Atlanta: Rodopi), pp. 203–30, on pp. 203–7.
60. The act provided three types of benefits; sickness benefit, statutory benefits (free access to GP services and specified drugs and appliances) and additional benefits (provided at the discretion of the individual insurance companies).
61. Ministry of Health (1924). *Reports on Public Health and Medical Subjects*, no. 23, *The Incidence of Rheumatic Diseases* (London: HMSO). See also D. Cantor (1991). 'The aches of industry: philanthropy and rheumatism in inter-war Britain', in J. Barry and C. Jones (eds). *Medicine and Charity before the Welfare State* (London: Routledge), pp. 225–45.
62. D. Cantor (1990). 'Rheumatism and the decline of the spa', in Porter, *Medical History of Waters and Spas*, pp. 127–44, note 5 on p. 28.
63. TNA: PRO. MH 55/1044, Royal Commission on National Health Insurance. Statement submitted on behalf of the United Patriot's National Benefit Society, p. 4.
64. Ibid., p. 9.
65. TNA: PRO. MH 62/35 'Clinics for the Treatment of Industrial Rheumatism', Henry Lesser (n.d. but c. December 1927).
66. TNA: PRO. MH 62/35 Ibid., noted as 'Extract from N.I. Gazette of 11 June 1927.
67. TNA: PRO. MH 62/35, Ibid., p. 3.
68. TNA: PRO. MH 55/1044, Royal Commission on National Health Insurance. Statement submitted on behalf of the United Patriot's National Benefit Society, p. 4.
69. TNA: PRO. MH 55/1044, B S F Circular No. 1. *British Spa Federation: Scheme for Spa Treatment for Independent or Voluntary Members of Friendly Societies* (undated). N. A. MH 62/35. Letter from John Hatton, Secretary of the BSF to Ministry of Health, dated 23 February 1929.
70. TNA: PRO. MH 62/231, Letter from Controller, the Ministry of Health to John Hatton, British Spa Federation, 6 July 1939.
71. 'The relations between the state, the health resort and national health insurance', *Archives Medical Hydrology*, VII, 1., (January 1929), pp. 126–43, on p. 126.
72. TNA: PRO. LAB 2/1929. 'Spa treatment for post office servants', June 1927 and 'Special arrangements for treatment of members of staff at British spas', issued by Ministry of Labour Benevolent Fund, April 1928.
73. 'Health Resorts Conference: Buxton's New Clinic', *BMJ*, 4 May 1935, pp. 940–1.
74. B. Jennings (ed.) (1970). *A History of Harrogate & Knaresborough* (Huddersfield: Advertiser Press), p. 447–8.
75. British Medical Association (1951). *The Spa in Medical Practice* (London: British Medical Association), pp. 28–9.
76. M. Langham and C. Wells (2003). *A History of the Devonshire Royal Hospital at Buxton* (Leek: Churnet Valley Books), p. 113.

7
Rethinking the Post-War Hegemony of DDT: Insecticides Research and the British Colonial Empire

Sabine Clarke

7.1 Introduction

The historical literature on insecticides and tropical disease is focussed overwhelmingly on the global Malaria Eradication Programme launched by the World Health Organisation (WHO) in 1955. The dominance at the WHO of the view that an aggressive programme of eradication using insecticides was the only acceptable option in the fight against malaria has led to the notion of the post-war hegemony of DDT.[1] What this hegemony meant in practice, according to historians such as Randall Packard, was a significant decline in scientific research after 1940.[2] It is repeatedly asserted that the dominance of insecticide-based control programmes retarded the understanding of malaria both as a biological and public health event. Packard states 'The adoption of a global malaria eradication programme by the WHO eradicated malariologists'.[3] A distinction has been set up in the literature between research capable of producing nuanced understandings of tropical disease and the crude application of the technological quick fix in the form of DDT. The recurring theme in the narrative of DDT use by rich nations is technological hubris. Post-war interventions in the tropics are said to have been characterised by uncritical faith in the superior nature of Western technology and its transformative power, which is said in the case of DDT to have had its origins in the experiences of the Second World War.[4]

There are two main problems that arise from this picture of the use of insecticides to tackle tropical disease after 1945. The first issue is that the global campaign for the eradication of malaria launched by the WHO in the 1950s was not a campaign that included sub-Saharan Africa, as many studies have noted.[5] Often, however, when historians

133

make the claim that the dominant eradication paradigm after 1945 led to a regrettable decline in scientific research into malaria they do not discriminate between countries within the ambit of the WHO programme and those outside. They also use the eradication approach of the WHO as a key context when describing the application of DDT in African countries.[6] In reality we know surprisingly little about the extent to which insecticides were deployed in African nations both before and after independence.[7]

The second issue is the claim that the approach of the WHO was representative of development policies more generally after 1945.[8] One aim here is to show that British approaches to tackling tropical disease as part of development initiatives of the post-war period were of a very different character to those espoused by the WHO. The eradication of malaria or sleeping sickness was not declared a goal across Britain's colonial empire. In addition, rather than seeing a decline in scientific research, the period after 1940 was a high-water mark for research in Britain's colonies, with around 40 new institutions created for this activity between 1943 and 1952.[9]

The development plans conceived by rich nations are often described as both grandiose in scale and simplistic in approach. Accounts generally focus on Africa and tell how experts reduced the complex nature of African environments and communities to crude pictures more amenable to expert intervention, and paid scant attention to indigenous knowledge.[10] One aim here is to show that attitudes towards the use of science in tropical countries did not always amount to uncritical deployment of the technological quick fix or the reduction of complex ecological and socio-economic relationships into simplistic pictures. The relationship that was construed between science and development in the late colonial period was more varied and complicated than some accounts would have us believe. The post-war privileging of research in British colonial policies meant British scientists examined in detail the efficacy of synthetic insecticides in controlling tropical disease before any application was considered. These studies often worked to reveal the complicated interactions that existed between insects, man and the environment. The net effect of this evaluation of insecticidal interventions, along with limits on the resources available to colonial states under British control, was not the rapid and ill-conceived application of new technology. Instead, researchers advocated caution and the results of their work tended to act as a check on any ambitions that may have been held by colonial officers to undertake the mass application of DDT in Africa.

7.2 From military deployment to colonial research

One key condition that shaped the character of British policies for the use of insecticides in mainland Africa was the emphasis placed on extensive scientific research into the new chemicals before their deployment as agents of disease control. Researchers from a number of disciplines and research institutions were mobilised for this work, which included medical research to assess toxicity, entomological studies into the behaviour of mosquitoes, engineering studies to develop spraying apparatus and chemical and meteorological research. What conditions led to the prioritising of research and allowed for novel arrangements in the post-war period to fund and oversee this multi-disciplinary research effort? Two factors can be seen to be important: the creation of administrative apparatus to oversee research into DDT during the Second World War that transcended any distinction between the civil and the military sphere and a strongly technocratic turn in colonial policy after 1940 which resulted in a substantial new source of funding specifically for scientific research.

Prior to the Second World War one of the main agents used in insect control was pyrethrum, derived from flowers of the genus chrysanthemum. During the 1930s pyrethrum was principally imported from Japan but with the entry of this country into the war, supplies to the Allied forces ended. In 1942 an Inter-Departmental Committee headed by Ian Heilbron from the Imperial College of Science and Technology, and a Research and Development Panel, were formed at the Ministry of Production in Britain to consider ways to meet the rapidly expanding demand for insecticides by the military.[11] Malaria and typhus were considered to present a significant threat to Allied troops stationed abroad with the potential, along with other diseases such as dysentery, to cause higher numbers of casualties than enemy engagement. In 1942 the Geigy Colour Company of Manchester released samples to the government of a new insecticide, DDT, which had been developed by its parent company, J. R. Geigy in Switzerland. In November 1943 the existing committees at the Ministry of Production became the Inter-Departmental Co-ordinating Committee on Insecticides and the Insecticides Development Panel and these bodies oversaw a programme of investigation into DDT. Research into the new insecticide was undertaken at a number of institutions including the Imperial College of Science and Technology, the Pest Infestation Laboratory of the Department of Scientific and Industrial Research (DSIR) and the London School of Hygiene and Tropical Medicine (LSHTM), where it was shown that DDT was a powerful agent against human lice, the carriers of typhus. Studies into the toxic effects of DDT on humans and animals were undertaken in Britain at the Chemical Defense Experimental Station at Porton Down in Wiltshire.[12] In 1944 a team from Porton Down travelled to Takoradi in the Gold Coast to undertake trials of DDT spraying by aircraft against adult and larval mosquitoes. Accompanying the team of chemists, physicists and engineers from Porton Down was the medical entomologist P. A. Buxton, based at LSHTM, who had previously investigated the effects of DDT against lice. The trials at Takoradi were intended to be a preliminary to the use of DDT in South-East Asia with the aim of determining the value of aerial spraying of areas forward of Allied lines to provide temporary relief from malaria when troops advanced.[13] Research of this sort led to the rapid deployment of DDT and aside from its use to control malaria during the Allied campaign in Burma and North-East India, DDT was used by Allied forces to control an outbreak of typhus among the citizens of Naples and troops stationed in Italy in early 1944. Buxton helped to develop shirts impregnated with DDT and these were given to British troops from 1944 as a measure to prevent infestation with body lice.[14] DDT was also used to control epidemic typhus in liberated concentration camps such as Belsen.[15] This particular episode was later used in an attempt to improve the image of DDT in light of what was portrayed by some as the disproportionate condemnation of synthetic insecticides in the early 1960s after the publication of Rachel Carson's *Silent Spring*.[16]

The success achieved with DDT during war time fuelled hope that a new era of animal and plant disease control by synthetic insecticides had dawned.[17] This ambition was given encouragement in Britain when the chemicals firm ICI announced in 1945 that they had developed a form of benzene hydrochloride (BHC), marketed under the name Gammexane, that was five times more potent against insects than DDT.[18] The experimental work into insecticides by state-funded institutions in Britain that had begun during the war continued after 1945 and the inter-departmental character of this work was retained. The committees set up during wartime to co-ordinate research into insecticides were replaced in May 1946 with an Insecticide Standing Conference and a Research and Development Co-ordinating Committee on Insecticides. The purpose of these two committees was to identify the research needs of government departments and bodies that had an interest in using insecticides and to relay these needs to the appropriate researchers, while making sure any results produced in the course of research were circulated to all interested parties. The committees brought together for discussion and collaboration representatives of the DSIR (who were concerned with the problems of pest infestation of stored produce), the Agricultural Research Council, the Medical Research Council and the majority of government departments including the Board of Trade, Ministry of Health, the Admiralty, Air Ministry and the Colonial Office.[19]

In March 1945 a request had come to the Colonial Office from the Governor of Uganda for staff, money and materials to undertake field trials of DDT and BHC against tsetse fly, which spread the parasite trypanosomes among animals and humans. Experiments were begun in Uganda at the end of 1945 with a small team, which studied both the application of insecticides to vegetation for tsetse control and the spraying of the interiors of houses to control mosquitoes. This team became the Colonial Insecticides Research Unit at Entebbe in Uganda headed by the medical entomologist C. B. Symes.

By January 1947 a new committee had been created at the Colonial Office to fund a comprehensive programme of research into insecticides and provide advice to colonial governments on the use of the new chemicals. The Colonial Insecticides Committee was one of ten committees created at the Colonial Office in London between 1942 and 1947 to deal with all fields of research considered to be relevant to colonial development.[20] A wide-ranging programme of experimental work was made possible because of the creation of a substantial Research Fund as part of the Colonial Development and Welfare (CDW) Act of 1940. With the act's renewal in 1945 the Research Fund had been increased to £1,000,000 per annum, making the Colonial Office the second largest sponsor of scientific research in the civil sphere in Britain in the period before 1950. At the Colonial Office, research was defined as work that examined fundamental issues in fields such as health or agriculture in the colonies and was said to underpin other more practical and problem-solving activities. The claim was made that while there had been steady increase in the numbers of technical officers in the colonies, particularly since World War I, there was not enough long-term research being done by scientific specialists to establish the basic facts about tropical environments. At the Colonial Office an emphasis on the acquisition of scientific knowledge through comprehensive research became an important element in a new and vigorous policy of improvement and modernisation for the colonies inaugurated with the passing of the 1940 CDW Act.[21] For staff of the Colonial Office in the early 1940's the promise of a substantial expansion in fundamental research was the provision of the means for effective planning and management of development.[22]

One effect of the emphasis on increasing both the amount and quality of scientific research in the British colonial empire was to bring large numbers of metropolitan scientists, often eminent ones, into the Colonial Office to create and oversee research projects. On its creation in 1947 the Colonial Insecticides Committee (CIC) comprised leading British experts in insect control, malaria and trypanosomiasis, a number of whom had been involved with the organisation and execution of insecticides research during the Second World War – the CIC was chaired by Ian Heilbron and P. A. Buxton also sat on the committee. Other members of the CIC included P. C. C. Garnham, Reader in Parasitology at LSHTM, D. L. Gunn, Director of the Anti-Locust Research Centre, W. J. Hall, Director of the Imperial Institute for Entomology at the Natural History Museum in South Kensington, George MacDonald, Director of the Ross Institute of Tropical Hygiene, J. W. Munro, Professor of Entomology and Applied Zoology at Imperial College and J. K. Thompson of the Colonial Office's Tsetse Fly and Trypanosomiasis Committee. Symes was appointed as Officer-in-Charge of Research with responsibility for determining the direction of insecticides research.

The main goal of the CIC was to evaluate critically synthetic insecticides through extensive research, as a preliminary to any large-scale trial or campaign in the colonies. The scientists in London and researchers in East Africa were to offer advice and practical help to any colonial government contemplating the use of the chemicals. The requirements of insecticides research in the British colonial empire were relayed to the inter-departmental Research and Development Committee on Insecticides. The CIC was represented on this body by the chemist John Simonsen who was appointed chair of the Research and Development Committee and R. A. E. Galley who was the secretary.[23] The high profile on the Research and Development Committee of individuals concerned with insecticide use in the tropics reflected the priority given to this particular application of synthetic insecticides. Galley defined the most pressing problem requiring the attention of British-based researchers as the development of effective spraying techniques, for use on the ground and also by aircraft.[24] Continuing an arrangement that had been established during the Second World War, the study of aerosols and the development of new spraying equipment was taken up by the Ministry of Supply's Chemical Warfare Defense Establishment at Porton Down. In the words of S. A. Mumford, who became Chief Superintendant at Porton Down in 1951, the work of the Chemical Warfare Defense Establishment scientists was concerned with, 'the use and adaptation of Chemical Warfare techniques for its employment in the control of the insect vectors'.[25]

In addition to the work at Porton Down there were field trials of insecticides in the colonies themselves. The centre of insecticides research in the British colonies was moved to Arusha, Tanganyika (now Tanzania) in 1950 with the creation of a new Colonial Insecticides Research Unit to replace the unit at Entebbe in Uganda. At Arusha a team of entomologists, chemists and physicists led by Kay Hocking did extensive work with experimental huts made of brick with thatched roofs and coated inside with mud to reproduce the interiors of African huts. In the field of tsetse control, the focus of experiments shifted from spraying the bush from the ground to using insecticides dispersed by aircraft and helicopters. Malaria and trypanosomiasis were priorities in research more generally at the Colonial Office as they were considered to be major obstacles to colonial development. Much attention was devoted to malaria by the Colonial Medical Research Committee that had been created by the Colonial Office in 1945 and which formed an East African Malaria Unit in Tanganyika in 1950. Three members of the CIC, Buxton, Garnham and MacDonald, sat on the Colonial Medical Research Committee. In a review of two decades of colonial research published in 1964, it was said that the development of DDT and drugs such as mepacrine was not considered to have removed the need for research into the epidemiology of malaria or the behaviour of mosquitoes. Rather 'the need constantly to attain a better understanding of that disease remained paramount, if counter-measures were to be used intelligently'.[26] CDW funds were used to deploy greater numbers of research workers in malaria to the colonies, to support research by colonial government malaria units and to increase laboratory studies at British institutions.

Similarly the field of trypanosomiasis research received an unprecedented level of support from the new Research Fund and investigations were overseen by a Tsetse Fly and Trypanosomiasis Research Committee created in London in 1944 and including Heilbron and Garnham among its members. Of the total spent on colonial research schemes by 1964 it was estimated that 8 per cent, or £1,920,000 was allocated to the field of tsetse and trypanosomiasis research. A West African Institute for Tsetse Fly and Trypanosomiasis was created in 1946 and a unit for East Africa was formed in 1947. The CIC also collaborated with researchers in the field of locust control.

Aside from the close relationship between research of military and civilian value one other legacy of the war for colonial insecticides research was regular contact with firms such as ICI. The team in Uganda had first received help from ICI when the company supplied a gas canister free of charge to produce insecticide smoke for treatment of the bush. After some discussion among members of the CIC it was decided that further collaboration with industry was acceptable on the condition that the CIC retained authority over the direction of research programmes and that the results of research were freely published.[27] The result was that scientists from Shell, ICI, Geigy and others worked with the insecticides' researchers on developing new formulations of chemical insecticides, and on field trials in East Africa of these preparations, throughout the period of investigation sponsored by the CIC.[28]

This comprehensive programme of research into the use of chemical insecticides to control tropical disease was made possible after 1945 because of a new colonial research fund of unprecedented size. It had been created as part of a reform of colonial policy that resulted in the passing of the 1940 Colonial Development and Welfare Act. The Colonial Office in London was keen to see high-calibre scientific researchers take a role in colonial development and believed that research would ensure that development planning in the post-war period would be based on sure knowledge of tropical conditions. The centre for insecticides research in the British colonial empire was established as the Colonial Insecticides Research Unit in Tanganyika. The field work done at this institution was co-ordinated with research institutions in Britain by London-based committees and was particularly closely related to laboratory studies carried out at Britain's chemical warfare research centre at Porton Down. This arrangement was a legacy of the mobilisation of scientific researchers to investigate chemical insecticides for use by the military during the Second World War.

7.3   Experimental work against tsetse fly in East Africa

One goal of British colonial policy was to increase the productivity of the colonies and the presence of tsetse fly, which made land uninhabitable for man and cattle, was considered a major obstacle to economic progress in its African territories. The reclamation of land infested with tsetse fly was a particular priority for British officers concerned with the denudation of existing land through intensive settlement and farming. The problem of tsetse was tackled by cutting down bush infested with the fly and then forcing Africans to resettle these areas in order to try and keep the bush and the fly at bay.[29] These policies could involve a degree of compulsion for both bush clearances and resettlement and were, unsurprisingly, often very unpopular with African farmers. The use of insecticides to eliminate tsetse fly seemed to offer a potential alternative tool for development officers and became, along with the control of malaria, a key area of insecticides research after 1945.

Researchers initially focussed on achieving adequate penetration of the bush by using aircraft to disperse synthetic insecticides. Two main issues arose from subsequent experiments in aerial spraying to control tsetse. One was the question of whether a safe level of tsetse fly, at which there was no further spread of trypanosomiasis, could ever be determined, or maintained. The other issue was the economics of disease control by insecticides. Reports on field experiments were frequently accompanied by a breakdown of the costs involved and by 1949 the committee in London was debating whether or not it was worthwhile continuing with aircraft experiments when the expense of this method of disease control seemed so prohibitively high.

An Aircraft Sub-committee, headed by Buxton who had worked on the wartime Gold Coast trials, was created at the very first meeting of the CIC to begin discussion of plans for experiments in the use of insecticides dispersed using aircraft and helicopters. At the sixth meeting of the CIC in September 1947 P. J. du Toit, Director of the Veterinary Research Institute at Onderstepoort, South Africa much impressed the scientists present with a showing of a film titled *Nagana*.[30] This film showed how the South African Air Force had sprayed 100 square miles of Zululand with DDT smoke to rid it of tsetse fly. The trial had been considered a huge success and it was suggested this method could be substantially more cost effective than the bush clearance methods that were normally used to make land safe for habitation and farming.[31]

The CIC made a request to the Treasury for aircraft to be based in East Africa for insecticides work and also planned to acquire a helicopter which could fly at lower altitudes and which would produce a downdraft to gave greater penetration of the bush. The case for helicopters was strengthened in the view of members of the CIC by reports that American firms were increasingly using them for crop and orchard spraying. In 1947 the committee were shown a letter that had been sent to the Colonial Office by Group Captain G. V. Howard. During his service with the RAF in Madras in 1945 Howard had been deeply impressed by a demonstration of insecticide spraying with an American Sikorsky helicopter. He was now planning to set up a company in Britain for crop spraying with helicopters and was applying for an import licence for three models from America.[32] All the evidence seemed to suggest that the future of tsetse control lay in the aerial application of chemical insecticides.

By May 1948 two Anson aircraft with crews had been chartered from Airwork Ltd and were based in East Africa. Engineers at Porton Down designed tanks to be fitted to the aircraft for liquid spraying and experiments were also carried out using smoke equipment fitted to the exhausts of the aircraft, based on that used in the South African project. Spraying experiments with aircraft were carried out across blocks of tsetse infested bush in Uganda and Tanganyika.[33] Scientists based at the Colonial Insecticides Research Unit at Arusha sampled the blocks after spraying to determine the reduction in levels of tsetse, returning again at a later stage to see if reinfestation had occurred.

As early as July 1949 concerns were raised by the committee in London about the aircraft experiments in East Africa, since retaining four pilots and four engineers from Airwork Ltd was very expensive. The aircraft trials were justified by scientists on the CIC by the claim that the costs of aircraft control would be offset by the increased value of the land that was cleared of tsetse and therefore available for the use of cattle.[34] Trials continued against tsetse infested blocks in East Africa but in January 1951 the question was raised at a meeting of the CIC as to whether tsetse fly were increasing to their previous levels in blocks of bush after treatment with DDT or BHC had finished. If treated blocks required continued further application by insecticides then the costs of this approach were even greater than first anticipated.

Officials at the Colonial Office started to issue warnings to the CIC in 1951 that it was going to have to prove more thoroughly the value of continuing with aircraft trials. The committee was warned to rein in its spending since the total allocation available for insecticides research for the five years between 1951 and 1956 was only £550,000.[35] This cast considerable doubt on the likelihood of the Aircraft Sub-committee achieving its goal of acquiring a helicopter for trials in East Africa, the cost of which was projected to be £100,000 over two years. Aside from the high costs of the aerial experiments, officials from the Colonial Office also raised their concerns about the inconclusive nature of the experiments so far when it came to demonstrating the effectiveness of DDT or BHC in the control of tsetse fly. A growing sense that synthetic insecticides were not proving to be a decisive means of tsetse control was shared by some of the scientists attending meetings. Gunn stated in 1951 that complete tsetse eradication was not possible through the aerial spraying of insecticides. The question then emerged as to what was a sufficiently low level of the fly to prevent the spread of trypano-somiasis. Some members of the CIC claimed experiments which pro-duced a 98 per cent reduction of tsetse could be said to be a success. This was disputed by H. M. O. Lester, head of the East African Tsetse and Trypanosomiasis Research Organisation, who stated that the deter-mination of a safe level of fly reduction was virtually impossible, since reinfestation could never be completely avoided. In addition, Lester informed the committee that experiments carried out in one locale could not be used as a basis for control in another as the conditions that led to the spread of trypanosomiasis were complex and the veg-etation and topography of an area affected the success of insecticidal treatment.[36] Lester's comments brought the future of any further experi-ments into question, since data produced in the course of one particular trial appeared to have no real value in determining the potential effi-cacy of insecticidal spraying elsewhere.

Those scientists that advocated continuing with aircraft trials referred again to the US experience in an attempt to show that success through aerial spraying of insecticides was possible, despite the inconclusive nature of the work in East Africa.

> It is not, perhaps, widely appreciated that in the USA some dozens of commercial firms are operating hundreds of aircraft for the control of pests and diseases on a variety of agricultural crops, fruit orchards and on forests, and that some American Government departments use aircraft for routine control of public health pests. It is notable, for instance, that after years of intensive study the Tennessee Valley Authority has fairly recently adopted aircraft application of mosquito larvicides as a routine method of control over very large areas of their impounded areas. It must be assured too that these air operations are accomplished for an economical cost for they would not continue if it were not so.

It was suggested that information be gathered on the American work and in 1953 D. Yeo, a physicist from the insecticides unit in Tanganyika, visited the US to study aerial spraying techniques.[37]

With no final decision made on the future of the aircraft in East Africa a series of experiments were arranged in an attempt to determine if a reduction in levels of tsetse was sufficient to eliminate trypanosomiasis in cattle. To ease the financial situation of the CIC the committee sought contributions from East African governments. Financial support from colonial governments was forthcoming as insecticides were considered to have potential as a tool for development. The hope was that DDT or BHC could be used to eliminate tsetse and free land for resettlement as part of schemes such as the Masai Development Plan launched by the Tanganyikan government in 1951.[38] This plan was intended to deliver new areas of grazing for Masai cattle and provide improved water supplies to Masai lands.[39] The Tanganyikan Government provided funds for experiments to investigate a potential role for aerial spraying as an aid to bush reclamation and experimental blocks were sprayed by the Colonial Insecticides Research Unit with DDT and gammexane.

The Government of Uganda also proposed a tsetse elimination experiment in the Lango district of the country. The spraying of 16 square miles at Maruzi with DDT in a gasoline and kerosene mix took place in 1953. A fence was then erected by the Uganda Tsetse Control Dept in an attempt to keep out game, as wild animals could bring infestation back into cleared areas.[40] In 1954, however, the secretary for Agriculture and Natural Resources, Uganda, wrote to the CIC to say that 'in spite of the considerable defence by way of pickets on the western side of the sprayed area, the fly started to seep back in'. The Uganda government informed the CIC that it would not finance further treatments of areas in Maruzi-Kwamia because of the extremely high costs, estimated at £225,000 for the entire area. The government also noted that even if the levels of tsetse were greatly reduced then the danger remained of a gradual return of the fly. 'Our conclusion is that spraying cannot by itself be considered to give a permanent answer to the problem of clearing an area of fly unless it is carried out in areas where geographical features provide natural defence against reinfestation'. The CIC was forced to concede that using insecticides from aircraft to control tsetse was likely to be neither practical nor economic.[41] In 1954 the contract with Airwork Ltd for the Anson aircraft was terminated and the Secretary of State for the Colonies instructed the CIC to disband the Aircraft Trials Sub-committee.[42]

The issue for both the Colonial Office and East African governments was that research revealed that application of DDT or BHC would not readily provide permanent control of tsetse. Control in the long term would likely require a regime of spraying and monitoring of the fly over a lengthy period in order to be certain that levels of tsetse remained too low for transmission of trypanosomes; a regime that would vary according to the environmental conditions of the site. The expense of using aircraft for the repeated application of insecticides was a major deterrent to continuing research into this method of tsetse control; the work was finally abandoned when it became clear that this was not a burden that colonial governments were willing to embrace. Research into insecticides revealed that rather than providing a simple quick fix, the effective use of this technology could potentially require considerable commitment of manpower, time and resources.

7.4 Experimental work in mosquito control

As with work in tsetse control, the attitude of the CIC towards the prospect of malaria control by insecticides was that there should be extensive research before any recommendations were made to colonial governments. Here also, the results of the research undertaken by the CIC into mosquito control only served to undermine any ambitions that colonial governments may have had in Africa for widespread spraying campaigns to control malaria. Historians of development in the late colonial period have talked of the role technical experts played in legitimating intrusive and disruptive large-scale projects carried out in the African colonies.[43] In the case of insecticides, however, the work of technical experts more often acted as a check upon any ambitions for mass spraying of African villages and the African landscape.

A Malaria Sub-committee was created by the CIC in 1947 which included Buxton, Garnham, MacDonald and Symes among its members. In contrast to the spraying of rural areas of bush to control tsetse fly, experiments into malaria control were focussed on the use of insecticides inside homes and other buildings. It was clear from an early stage, however, that the use of insecticides to control malaria in Africa was not going to be straightforward. The first annual report of the CIC stated that a substantial number of investigations would be required before the widespread use of insecticides as tool for malaria control in the tropics could be contemplated. This report and subsequent ones did not set out bold claims about the eradication of malaria or trypanosomiasis in East Africa. The CIC were involved, however, in one large-scale project with the aim of malaria eradication and this was undertaken on the island of Mauritius. The costs of this large-scale eradication programme were borne by the Mauritian government, who applied for a CDW grant while the CIC provided research personnel.[44] The scheme had been inspired by the *Anopheles* (Malaria) Eradication Campaign in Cyprus, which had been launched in 1946.[45] This project was also funded by a CDW grant from the British government and was carried out by the Cyprus Sanitary Department.

In 1947 the director of medical services in Mauritius, Dr A. Rankine, attended a meeting of the CIC and managed to secure its help for a campaign to eradicate malaria using insecticides. This began in January 1949 and consisted of island-wide spraying with DDT solution of the interiors of all dwellings. The preparation for this scheme included surveys of the countryside and the identification of all houses and other buildings. The inhabitants of Mauritius found they had no choice when it came to participation, since access to homes for spraying and data collection was guaranteed by the introduction of a government ordinance in 1948 that gave staff employed on the campaign the authority to enter all buildings.[46] Mauritius was eventually declared free from malaria in 1952 and the scheme was followed by a study of mosquito ecology by an entomologist sponsored by the CIC.

The focus of the experimental work at Porton Down and East Africa funded through the CIC was on the determination of the optimum formulations, and best application regime, of the various insecticides available in order to substantially reduce mosquito presence in built up areas. The main tool for the investigations in the field was the experimental hut. Insecticides were applied to the interior of brick huts that had been coated with mud traditionally used in African housing and traps were set up to measure the levels of mosquitoes. Illustration 7.1 shows experimental huts at Magugu, 90 miles southwest of Arusha in Tanganyika; the picture was used by the Colonial Insecticides Research

Illustration 7.1 'Experimental huts at Magugu'; reproduced by kind permission of Kay S. Hocking

Unit to illustrate a report published in 1959. One key issue for the researchers in Tanganyika was to determine the most effective insecticide formulation from the increasing choice of available chemicals, with dieldrin and aldrin emerging alongside DDT and BHC by the 1950s. The main problem was to find a suitable medium for application of insecticides that prevented absorption by the mud coating, and therefore a loss of toxic effects. The effects of DDT in a solution or emulsion wore off very quickly when applied to the mud that coated the interior of huts in East Africa as the chemical was absorbed by the walls. Experiments done by the Colonial Insecticides Research Unit in Tanganyika found that wettable powders retained their toxicity to a far greater extent. The team in East Africa was also concerned with determining the frequency of reapplication of each insecticide formulation if low levels of mosquitoes were to be maintained.

This research in the field was supported by laboratory experiments at Porton Down in which researchers investigated the optimum size of insecticide crystals to ensure that the chemical was readily picked up by the feet of mosquitoes. Studies into the behaviour of mosquito species were also carried out to gain an understanding of their daily habits and feeding patterns. Laboratory studies in the UK were expanded from 1949 to Silwood Park run by Imperial College and the agricultural research centre, Rothamsted, in Hertfordshire.

In general the results of experiments in the field were very disappointing and failed to identify a cheap and effective long-term method of mosquito management. Discrete and small-scale studies very rarely progressed into schemes for large-scale malaria control and the larger projects into malaria control that were undertaken were considered failures. One two-year trial in a township in Uganda was undertaken in which the interiors of all houses were treated with wettable DDT powder. This trial was declared unsuccessful in 1951 since the incidence of malaria among residents showed no change, prompting the Colonial Office to comment: 'This is an important observation, since so many local authorities in Africa are anxious to adopt insecticidal measures in the many townships or small aggregations of populations'.[47] Similarly a project in Nigeria that ran from 1954 to 1956, the Western Sokoto Malaria Project, was described after a visit by Galley as showing very disappointing results. The project had failed to stop malaria transmission in its first phase and the second phase had revealed resistance to dieldrin by *Anopheles gambiae*. While the emergence of resistance has been said to have spurred the WHO to recommend rapid and widespread deployment of insecticides as part of its global malaria eradication programme[48] the response of the CIC was typical. For the CIC the news of insecticide-resistant strains of mosquito was confirmation that eradication in Africa was not an achievable goal in the foreseeable future and that the most pressing and urgent need was for more research.[49]

Between 1954 and 1959 the insecticides unit at Arusha participated in the large-scale Pare-Taveta trial organised by Donald Bagster-Wilson of the East African Institute of Malaria and Vector-Borne Diseases. This large-scale project has been the subject of some confused historical writing.[50] The Pare-Taveta Malaria Scheme was not an attempt at malaria eradication or a pilot trial as a precursor to any larger campaign but was designed by Bagster-Wilson as an experiment. It was sponsored by funds from the British CDW allocation and the East African governments and was neither principally funded nor directed by WHO or UNICEF as has been claimed, although UNICEF provided the insecticides.[51] Michael Gillies who worked as a medical entomologist on the trial has written that it was disagreement with the WHO's emphasis on eradication of malaria by insecticides that prompted Bagster-Wilson to undertake the experiment. By the time the Pare-Taveta trial began Bagster-Wilson had retired from the WHO Expert Committee on Malaria because of frustration with the dogmatic promotion of eradication by his colleague, George MacDonald.[52]

Bagster-Wilson wished to examine the impact of malaria on the health of an African population. He believed that intensive use of insecticides to control malaria would only work to reduce immunity and he was concerned that the WHO approach meant the neglect of the other myriad diseases that contributed to poor health in African communities.[53] The main objective of the study was to interrupt the transmission of malaria for a period and then survey the African population in order to ascertain the burden placed upon health and fertility by malaria in relation to other common diseases.[54]

The trial lasted between 1954 and 1959 and involved the spraying of around 17,500 homes with dieldrin on six occasions. The experimental area ran approximately 100 miles in length from the Pare District of Tanganyika to Taveta in Kenya and consisted of low-lying places of endemic malaria and other areas of higher ground. The involvement of the Tanganyikan insecticides research unit was aimed at making an assessment of the efficacy of dieldrin in killing mosquitoes, with large numbers of traps set up to assess the mosquito population before and after insecticide application. Human mosquito catchers were also employed to sit outdoors and catch insects that alighted on their legs in the evening to determine the extent to which bites from mosquitoes outside of homes could potentially spread malaria.[55] The results of this work showed that despite the multiple rounds of spraying the incidence of malaria never fell to zero, most likely because a small proportion of mosquitoes avoided the insecticidal treatment by resting outside. Gillies wrote in his memoirs that the trial exposed the paradox of insecticidal spraying campaigns to control malaria in rural Africa. The persistence of malaria meant that once spraying had been discontinued it was only a matter of time before the incidence of the disease rose to its previous levels. The high expense of continued spraying, however, was not justifiable since, 'there was the ever present possibility that sooner or later the mosquitoes would become resistant to the insecticide and the whole enterprise would come to nothing'.[56]

Medical researchers on the Pare-Taveta project undertook examinations of large numbers of people during which their heights and weights were measured, blood was drawn and the liver and spleen were palpated. Signs of improved general health and fertility in the absence of malaria were difficult to find and the results of the experiment were considered rather inconclusive. Any improvements in health and fertility were difficult to attribute to a temporary absence of malaria since the trial had coincided with a period of increased prosperity in the region and any gains measured could well have been related to better nutrition. Bagster-Wilson's trial had served only to expose the complex relationship that existed between the health of a population and a range of social, economic and environmental factors.[57]

Literature that considers the role of scientists in the European empires after 1945 has told of a simplifying and generalising impulse at work in the representations of colonial environments produced by these experts, and the interventions they conceived.[58] In contrast, entomologists, chemists and medical researchers working with the new chemical insecticides such as DDT contributed to increasing complex and nuanced understandings of tropical disease and environments in the course of their investigations. Rather than simplifying and generalising, the tendency of this multi-disciplinary research effort was more often to complicate existing understandings and to expose the shortcomings of disease-control methods based exclusively on the use of synthetic insecticides.

7.5 Conclusion

In recent years there has been increasing scholarly interest in the importance given to technical expertise when it came to interventions in tropical nations by the West after 1945. The picture given is often one where faith in Western science and technology led to simplistic, ill-conceived and sometimes highly disruptive large-scale development schemes, which produced few positive results. Histories of the WHO global campaign to eradicate malaria claim tend to conform to this general pattern, stating that dependence on the technological quick fix of DDT displaced other more sophisticated and responsive approaches to the control of tropical disease after 1945. The relationship between science and development in the post-war period was more varied and complex, however, than many of these stories would have us believe. There was increasing diversity in the type of scientific activity undertaken, and of practitioner deployed. Alongside such things as large-scale agricultural schemes led by technical staff who advised on farming practices, we also find in the British case the substantial deployment of scientific specialists after 1940 sent to the colonies to undertake detailed studies of such things as tropical diseases, soils, forests and fisheries. Insecticides research in British East Africa was bolstered by the interest that DDT and similar compounds received at home, benefiting from arrangements put into place during the Second World War for the co-ordination of research among different institutions, including those of the military. Entomologists were particularly well represented in the research committee and units established to carry out research and there were close links with medical researchers and institutions such as the London School of Hygiene and Tropical Medicine. A comprehensive programme of research was established after 1947, which included the study of insect behaviour and malaria epidemiology along with chemical and engineering studies. In the British colonial empire, the new technology of DDT did not supplant studies of tropical disease and insect vectors. Research involving the new chemical insecticides contributed to more sophisticated understandings of the interactions between the diseases of animals and man and the environment in Africa.

The privileging of scientific research in British colonial policy after 1940 meant that before deployment of insecticides in Britain's mainland African colonies was considered, researchers were appointed to assess the efficacy of DDT and other chemicals against insect vectors of disease. Comprehensive experiments in the laboratory and the field revealed two main problems with deployment of insecticides as a means of disease control. One was that insecticidal treatments did not offer a permanent solution to the spread of disease by mosquitoes or tsetse fly. The second was that colonial governments considered the deployment of insecticides to be prohibitively expensive. The net effect of the work of scientific researchers was not, therefore, to sanction any simple, technical solutions to colonial problems but to act as a check upon ambitions for large-scale deployment of chemicals for the control of disease in Britain's African colonies. DDT and similar chemicals were shown to be of little value in providing a demonstration of the seriousness of Britain's commitment to colonial development after 1940.

Notes

1. See articles in the special issue of *Parassitologia, Strategies against Malaria: Eradication or Control*, 40 (1998), especially J. Jackson, 'Cognition and the global malaria eradication programme', pp. 193–216 and R. Packard, '"No other logical choice": global malaria eradication and the politics of international health in the post-war era', pp. 217–29; R. Packard (2007). *The Making of a Tropical Disease: A Short History of Malaria* (Baltimore: John Hopkins University Press); J. L. A. Webb (2009). *Humanity's Burden: A Global History of Malaria* (Cambridge: Cambridge University Press); M. J. Dobson, M. Malowany and R. M. Snow (2000). 'Malaria control in East Africa: the Kampala conference and the Pare-Taveta scheme: a meeting of common and high ground', *Parassitologia*, 42, pp. 149–66; L. Schumaker (2000). 'Malaria', in R. Cooter and J. Pickstone (eds). *Medicine in the Twentieth Century* (Amsterdam: Harwood Academic), pp. 703–17.
2. Jackson, 'Cognition'; Packard, '"No other logical choice"'; Packard, *The Making of a Tropical Disease: A Short History of Malaria.*
3. Packard, '"No other logical choice"' p. 219.
4. Jackson, 'Cognition'; Packard, '"No other logical choice"'.
5. Schumaker, "Malaria", pp. 703–717; Jackson, 'Cognition'.
6. Dobson, Malowany and Snow, 'Malaria control in East Africa'; Schumaker, 'Malaria'; D. J. Bradley (1998). 'Specificity and verticality in the history of malaria control', *Parassitologia*, 40, 9; Packard, *Making of a Tropical Disease*; J. McGregor and T. Ranger (2000). 'Displacement and disease: epidemics and ideas about malaria and Matabeleland, Zimbabwe, 1945–1996', *Past and Present*, 167, pp. 203–37.
7. One exception being McGregor and Ranger, 'Displacement and disease'.
8. Packard, *The Making of a Tropical Disease*, pp. 144–5.
9. S. Clarke (2007). 'A technocratic imperial state? The colonial office and scientific research, 1940–1960', *Twentieth-Century British History*, 18, pp. 453–80.
10. Packard, '"No other logical choice"'; C. Bonneuil (2000). 'Development as experiment: state and state building in late Colonial and postcolonial Africa, 1930–1970', *Osiris*, 15, pp. 258–81; J. McCracken (1982). 'Experts and expertise in colonial Malawi', *African Affairs*, 81, pp. 101–16. More recent work has noted the caution that could be advocated by European experts when it came to intervening in areas such as African agriculture, see J. Hodge (2007). *Triumph of the Expert: Agrarian Doctrines of Development and the Legacies of British Colonialism* (Athens: Ohio University Press).
11. National Archives of the UK (NA), MAF 117/23.
12. I. M. Heilbron (1945). 'The new insecticidal material, DDT', *Journal of the Society of the Arts*, xciii, pp. 66–7.

13. NA, WO 203/158; WO 189/2576.
14. NA, WO 188/739.
15. M. Harrison (2007). *Medicine and Victory: British Military Medicine in the Second World War* (Oxford: Oxford University Press), p. 270.
16. J. Sheail (1985). *Pesticides and Nature Conservation: the British Experience 1950–1970* (Oxford: Clarendon Press), p. 93.
17. Heilbron 'The new insecticidal material, DDT'; It is important not to over-state the contribution to disease management amongst allied troops made by DDT in comparison to sanitary work and chemical prophylaxis, see Harrison, *Medicine and Victory*.
18. W. J. Reader (1975). *Imperial Chemical Industries: A History*, vol. 2 (Oxford: Oxford University Press), pp. 454–7. Gammexane was also known as 666.
19. NA, MAF 117/23.
20. Clarke, 'A technocratic imperial state?'.
21. Ibid.
22. Ibid.
23. Robert Albert Ernest Galley had been Principal Experimental Officer in the Chemical Inspectorate, Ministry of Supply during the war, becoming Senior Principal Scientific Officer at the Agricultural Research Council in 1946. In 1960 he became a Director of Shell Research Ltd.
24. NA, MAF 117/23.
25. NA, WO 188/739.
26. C. Jeffries (1964). *A Review of Colonial Research, 1940–1960* (London: HMSO), p. 91.
27. NA, CO 911/1.
28. *Colonial Research, 1950–1951*, Cmd 8303.
29. D. L. Hodgson (2000). 'Taking stock: state control, ethnic identity and pas-toralist development in Tanganyika, 1948–1958', *Journal of African History*, 41, pp. 55–78; Hodge, *Triumph of the Expert*, pp. 214–16; McCracken, 'Experts and expertise in colonial Malawi'.
30. On the history of the laboratory at Onderstepoort see K. Brown (2005). 'Tropical medicine and animal diseases: Onderstepoort and the development of veterinary science in South Africa, 1908–1950', *Journal of Southern African Studies*, 31, pp. 513–29.
31. NA, CO 911/1.
32. Ibid.
33. NA, CO 927/141/6.
34. NA, CO 911/3.
35. NA, CO 911/5.
36. Ibid.
37. NA, CO 911/6.
38. NA, CO 911/5.
39. For details of the Masai Development Plan see Hodgson 'Taking stock'.
40. NA, CO 911/8.
41. Ibid.
42. Ibid.
43. See for example Hodge, *Triumph of the Expert*; Hodgson 'Taking stock'.
44. *Colonial Research, 1947–1948*, Cmd 7493.

45. K. Constantinou (1998). '*Anopheles* (malaria) eradication in Cyprus', *Parassitologia*, 40, pp. 131–5. The scheme was formally declared a success in 1950.
46. M. A. C. Dowling (1953). 'Control of malaria in Mauritius', *Transactions of the Royal Society of Tropical Medicine and Hygiene*, 47, pp. 177–98.
47. *The Colonial Territories, 1950–1951*, Cmd 8243.
48. Packard, '"No other logical choice"', pp. 217–29.
49. NA CO 911/10.
50. Dobson, Malowany and Snow 'Malaria control in East Africa'.
51. Ibid., p. 163.
52. M. Gillies (2000). *Mayfly on the Stream of Time* (Hamsey: Messuage Books), p. 171.
53. East African Institute of Malaria and Vector-Borne Diseases in Collaboration with the Colonial Pesticides Research Unit (1960). *Report on the Pare-Taveta Malaria Scheme 1954–1959* (Dar es Salaam: Government Printer).
54. Ibid.
55. Ibid.
56. Gillies, *Mayfly on the Stream of Time*, p. 176.
57. *Report on the Pare-Taveta Malaria Scheme*.
58. Bonneuil 'Development as experiment'; D. Hodgson and M. Van Beusekom (2000). 'Lessons learned? Development experiences in the late colonial period', *Journal of African History*, 41, pp. 29–33.

8

'Health Crusades': Environmental Approaches as Public Health Strategies against Infections in Sanitary Propaganda Films, 1930–60

Christian Bonah

The control of contagious diseases has a long history. Since the late nineteenth century hygiene has been remodelled in the Western world according to the principles of bacteriology[1] transforming hygiene into a concern for public health and infectious diseases.[2] Nevertheless, engagement with environmental issues persisted even in the era of microbial causality.[3] The concept of linking isolation and total containment with the efficacy of bacteriological tracking and laboratory-administrative control of specific pathogenic microbes was thrown into doubt by the experience of the ubiquitous 1918 pandemic influenza, and which opened a new era of competing models of disease causation and new research practices such as experimental epidemiology.[4] For the interwar decades, infectious diseases were thus disturbances of a natural, hidden equilibrium between microbes, host populations and their surroundings. Cycles of transmission became more complex with animals conceived as intermediary hosts, disease reservoirs and vectors.

Hygiene was an applied science. The social engineering of human behaviour and administrative policing of individuals and populations were both tedious. There existed two complementary or alternative approaches: the biomedical laboratory-based and vaccine- or chemotherapy-oriented approach on the one hand, and, on the other, the environmental one, which instead focused on physically or chemically transforming landscapes and urban environments conceived as reservoirs and breeding spaces for microorganisms that cause infectious diseases.[5] Both were science- and technology-based.

The following contribution will analyse how infection was understood in three crucial decades before and after World War II. These decades can be characterised as being population-oriented, complexity-focused and often still lacking in curative therapy, which meant that to 'transform the environment' – from drying wetlands to spraying insecticides – remained one of a series of practical advisable strategies applied to promote healthier living. How this understanding was promoted and presented to professionals and the wider public will be approached through what the actors of the time considered the most 'modern' audiovisual techniques of persuasion and information: sanitary instruction and 'propaganda films'.[6]

The inception of motion pictures has been crucial in establishing relationships between science, medicine and society in the twentieth century.[7] Accordingly, films are a fascinating medium for studying the cultural production of scientific representations of 'reality' and their translation and integration into public perceptions of health and disease.[8] The present analysis is based on the classical distinction in film studies between fiction and non-fiction film and relies exclusively on the latter as sources. The rationale behind this choice is that non-fiction films are less frequently used as sources and they allegedly propose to present their subject as a 'realistic' translation of facts for their audiences.[9] Although in strictly chronological terms the first health-related motion pictures were developed in the pre-1914 era,[10] World War I and the interwar years witnessed an unparalleled development of sanitary educational and propaganda films. Rather than focus on an individual production, this contribution makes the methodological choice to review representations of the environment and health in a series of films over a longer period from the 1930s to the 1950s. As one possible approach, we propose to study a corpus of films according to their inherent historical development.

What characterises our selection of four films discussed in detail is that they were considered 'films of facts'.[11] They cover the entire 20-year period and represent sponsorship and production by four major film-commissioning institutions: the industry (Shell Film Unit), the government (US Public Health Service, USPHS), the army (US War Office) and finally, international organisations (United Nations, UN, and World Health Organisation, WHO). Our choice concerns exclusively productions from the Western world and thus portrays a North American and European point of view. Films selected are not strictly speaking environmental films by subject but rather films representing major categories of mainstream health education and training films. They are used to assess the role of environmental approaches within the wider setting of propaganda-mobilising citizens in national and transnational efforts to alleviate and control infectious diseases.

Chronological limits of the presentation extend from early projects of environmental interventions in the 1930s to the 1950s. Equal attention is given to World War II films and the post-1945 era, not only because World War II was a significant and autonomous period, with wartime conditions intensifying public health propaganda and actual film production, but also because it represented an intense moment of conceptual and technical change.[12] It will be argued that major changes were carried over from wartime film into post-war reorganisation. For purposes of distinction and to discuss more clearly the transition from the 1940s to the 1950s, films chosen during World War II are non-military films meaning here that they were not primarily concerned with the war and thus are comparable to non-fiction films before and after the war.[13] The chronological frame is extended beyond World War II, with the intention of questioning the systematic distinction between 'information' and 'propaganda' as these terms are traditionally accepted, and also to allow an analysis of the globalisation of public health after 1945. From a film history perspective, ending with the 1950s reflects the idea that this is a period of major transition, especially for 'factual films'. In the 1950s the audiovisual sphere was transformed with the appearance of television as a new method of film distribution, and new camera equipment and techniques (including synchronous recording of sight and sound) which led to Direct Cinema and *Cinema vérité*.[14]

Finally, the analysis is not centred on a specific disease. The films are rather analysed as autonomous visual entities of a given time, representing convictions and approaches to advocating disease control strategies aimed at professional or general audiences. They are thus mobilised here as fora for institutional advocacy and as illustrations of the ways in which these institutional players conceived intervention on the environmental factors which co-determined infectious disease control.

What evidently is missing in this filmic overview is more detailed information on the precise circumstances of production and distribution of the individual films presented and a more systematic inventory of the variety of films produced and projected. If individual films presented here have been chosen as representing a line of motion picture argumentation throughout the period, they nevertheless remain icons in an ocean of images. Further detailed contextualisation of the films through production records, correspondence between commissioning institutions and producers, audience reactions in film reviews or private testimony is evidently highly desirable, but beyond the scope of this presentation.

8.1   Origins of health education movie elements

Film is a heterogeneous and hybrid medium. Health education and teaching films had several origins and integrated sequences, elements and features common to many of them. First, after the inception of motion pictures during last decade of the nineteenth century, the use of cinematography in biological and medical research as recording devices experienced a period of intense growth between 1900 and 1914.[15] Most significant for the films analysed below, was the invention of micro-cinematography in 1907 by the French biologist Jean Comandon meaning that a film camera could be applied to a microscope in order to display living bacteria[16] visualising their vital existence.[17]

Second, early clinical instructional films appeared. For example in Boston, Walter G. Chase produced from 1905 onward motion pictures depicting epileptic seizures and similar clinical observations. Such clinical teaching films developed in hospitals throughout the Western world as short and to-the-point case report illustrations.

A third variety of medical cinematography involved the use of films in health education. These films started to be distributed during the first decade of the twentieth century in many European and North American countries. Beginning in 1910, a series of motion pictures were produced by Thomas Alva Edison for the National Association for the Study and Prevention of Tuberculosis in the United States.[18] Edison's films showed how, in the interests of public hygiene, the development of modern techniques of propaganda and mass media persuasion could be mobilised to promote strategies of disease control through 'public education' especially in efforts like the war on tuberculosis, venereal diseases and alcoholism. With the appearance of feature-length films during World War I, motion pictures were used, to varying degrees, by the armed forces for training, historical recording, public information and propaganda purposes.

The fourth origin of so-called 'educationals' was due to the initiative of film industrialists like Charles Pathé in France. They made possible the production of instructional films for schools, universities and for lay public education at a level of high cinematographic and scientific competence, and in particular developed animation techniques as pedagogical tools. Promoting the advancement of science and conceiving the development of serious and useful forms of entertainment, companies like Pathé offered their support to scientists-turned-film directors, as in the case of the earlier-mentioned Comandon. In 1909 Comandon became director of the newly created Scientific Service for Pathé, which gave him a free hand to produce educational motion pictures on biological and physiological subjects. Comandon produced the first film combining microscopic and close-up techniques dealing with the development of the mosquito from the egg to the hatching insect and the asphyxiation of the larva. On the German side in 1917 the Universal Film Aktiengesellschaft (UFA) was founded in Berlin as a common stock company with funds provided by the German government.[19] UFA promptly set up an Educational Department (*Kulturfilm-Abteilung*) which produced educational and health propaganda films throughout the period of Weimar Germany. In Great Britain, educationals were linked to the documentary film movement and in 1930 the establishment of the Empire Marketing Board (EMB) Film Unit by John Grierson. In 1933 the EMB Film Unit moved to the General Post Office and Grierson convinced the oil company Shell International to establish its own film unit under the direction of Edgar Anstey.[20] These developments formed the background against which non-fiction films concerned with health and environment presented here appeared.

8.2 Contents and characteristics of a selected film corpus depicting health and environment

Starting in the interwar years, industrial corporations were heavily engaged in the production of commercial films that intertwined educational scientific elements with advertisement and promotional propaganda.[21] Almost completely ignored by historians and film studies alike, commercial films became quantitatively extremely significant from the late 1920s onward and they are of particular interest for the history of environmental health studies as the following example will show.

Several industrial films dedicated to malaria[22] and its public health treatment were produced in the early 1930s. They integrated a long-standing effort of malaria public health education. If earliest films on malaria were produced since 1912 by commercial companies like Pathé, Gaumont, Edison and British Instructional Films (often included in newsreels), then agency-commissioned films devoted to the disease started in 1925, with a film produced by the Rockefeller Foundation, and peaked during World War II.[23] Besides a series of German malaria films[24] co-produced by UFA and Oskar Wagner, who in 1926 left the UFA to organise film production for the chemical and pharmaceutical company Bayer (Bayer Film Unit),[25] a significant and technically up-to-date example is the motion picture *Malaria* produced by the petrol giant Shell Film Unit.[26] The film selected for analysis is presumably the 1941 sound re-edition of an initial 1931 production under the same title. It was released in France during World War II.[27] The film is thus considered here as an example that illustrates the time span between 1931 and 1941. The film started as a purely scientific teaching film strongly resembling Comandon's pre- and post-World War I work. As the Bayer malaria film did, it integrated elaborate microscopic and close-up views and sophisticated animated drawings of the malaria pathogen Plasmodium and its reproductive cycle throughout the first section (ten minutes), thus integrating elements imported from earlier research and educational films. The second part of the film was dedicated to the intermediary host, the mosquito, and its living habits in a natural environment. The third section addressed protective measures applicable to mosquito-infested areas. Following the general claim that 40 years after the discovery of Plasmodium transmission by the mosquito by Ronald Ross – in 1897 – malaria was still extremely frequent and individual medical action was uncertain, the central message of the film implied that the fight against the mosquito should be a top priority for public health approaches. If individual treatments such as repellents and quinine were mentioned, in a global approach they were considered as just as limited as the traditional means of mosquito door and window screens, and mosquito nets. Environmental solutions were presented as cheaper and supposably more successful since their efficacy was independent of individual human compliance. The 'environmental approach' was a solution to sideline and replace conditioning human behaviour (e.g. instructions to cover all body parts with clothing, wear boots, how to tighten mosquito nets correctly, to apply repellents or to ingest quinine etc.) which was considered too tedious, unpredictable and uncontrollable.[28] Furthermore traditional physical-technical environmental solutions such as drainage were supposed to be superseded by even cheaper and more 'modern' solutions such as 'oil larviciding' (see Illustration 8.1).

Oil larviciding consisted in the systematic spraying of petrol on all water surfaces of a malaria-ridden area, thus inhibiting mosquito larvae from obtaining oxygen necessary for their development.[29] Evidently, the petrol used for this 'environmental treatment'[30] needed to be of best quality, which brings us back to the sponsor and producer of the film, the petrol company Shell (see Illustration 8.2). The concluding message of the film stated that 'man is often responsible for defective

Illustration 8.1   Oil-spraying operator. *Malaria* (1931/1941). Published courtesy of the DHVS-IRIST film collection, University of Strasbourg

engineering – a work of anti-social significance – and that oil larviciding will contribute significantly to enhance work and life in malaria-infested regions – especially in the African and Indian territories and colonies'.[31] If the Shell film was first and foremost a tool of corporate communication and instruction for employees and destined for colonial overseas collaborators, it was also used for distribution in schools, universities and to associations interested in documentary film and science.[32] *Malaria* included scenes in the British colonies including India, and North and Central Africa that depicted settler communities as well as local populations both being included in health efforts to prevent the disease even if the underlying tone was one of paternalistic settler superiority and white British engineers directing indigenous labourers (a group of men drying swamps and the individual worker spraying oil are all indigenous working populations (see Illustration 8.1).

If the 1930s can be portrayed as a period of steady, but unspectacular, growth of non-fiction medical films, the 1940s and especially World War II witnessed a steep rise in the production of wartime training films. Paradoxically, the increase of medical film production during this period

Illustration 8.2   Laboratory scientist of the Shell Corporation testing the quality of oil used for larviciding. *Malaria* (1931/1941). Published courtesy of the DHVS-IRIST film collection, University of Strasbourg

is inversely proportional to our knowledge about them. Considering that army films were important but at the same time represented a rather distinct category, due to their specific audience, the second film selected represents 'ordinary' government initiatives during wartime.

Ten years after first corporate malaria films were produced, oiling as a malaria control practice and film as tool for instruction became generally adopted to the point that the USPHS issued a 12-minute training film *Oil Larviciding (1943)* for larviciding operators.[33] The film straightforwardly explained that oil larviciding was an effective way to combat mosquito-born diseases by killing the aquatic stages of the vector. Diesel oil was presented as the preferred agent because it was highly toxic, not too volatile, and it left an even, stable film on the water surfaces treated (see Illustration 8.3). According to the training film advice, it was best applied in a spray. From there on the film presented different locations of the continental United States where malaria was endemic in the 1930s. Scenes of swampland from southern states indicated that

Illustration 8.3 Animation explaining the practice of oil larviciding and the quantities of oil needed to treat a given water surface. *Oil Larviciding* (1943). Published courtesy of the National Library of Medicine

small areas needed to be hand sprayed; larger rural areas could be power sprayed from boats (see Illustration 8.4) and trucks, and motorcycles could be used in cities. The film made extensive use of animation (see Illustration 8.3), explaining technical details of oiling operations, showing how the oil film on the water killed the larvae and pupae, and presenting different apparatus.

The film can thus be viewed as a 'homefront' movie participating in the national mobilisation against diseases that gained actuality through the war but at the same time it represents rather apolitical and factual information unrelated to the war itself. From a conceptual point of view the film was a technical instruction movie for workers employed as oiling operators, but again the film was used for public education in schools as well. If the film displayed many of the technical devices and the sophistication of the war propaganda movement, at the same time it illustrated the generalisation of film even for low priority activities such as instructing oiling operators.

Illustration 8.4 'Power oiling' using a boat to spray large water surfaces. *Oil Larviciding* (1943). Published courtesy of the National Library of Medicine

After 1945, films produced during the war were used in research, education, and training. Continuities abounded as American film productions became part of post-war reconstruction efforts by the US Information Service (USIS) and many films were translated into European languages, including Disney wartime productions.[34] The UN Educational, Scientific and Cultural Organisation (UNESCO) put forward a world plan of work for the mass media – films, radio and press – in 1947.[35] The programme produced a survey of films and other direct services in 1947, eventually becoming the UN Film Board, responsible for coordinating all the film work of the UN's special agencies, especially the Food and Agriculture Organisation (FAO) and the WHO. World War II thus directly triggered post-war organisation and influenced health information campaigns that were waged immediately after the war by national and international organisations. Two examples of the early 1950s illustrate the focus of these efforts.

Practical Rat Control. Rat Killing, a film produced in 1950 by the US War Office with the advice and assistance of the Communicable

Disease Centre in Atlanta and the USPHS as well as the Federal Security Agency, depicted on-screen US rat control efforts.[36] The black-and-white 12-minute film was part of a larger project on rat control that started in 1942 with the *Keep them out* production.[37] US War Office and PHS efforts to combat rat infestation of major cities intensified in 1950, with a complete film series on rat control.[38] *Practical Rat Control. Rat Killing* showed rats' habitats and their destructive and disease-spreading activities. The film was centred on rat elimination and control measures to be performed by the community. The film was intended to alert public consciousness about nuisances caused by rats in professional and domestic premises, and to detail organisation of dispositions to be taken to eliminate rats as disease carriers. The film started with an explosion and war scenes and the commentator pointed out that the US knew how to defend itself against human invaders. The off-voice continued 'but there is another type of enemy' and the camera presents a close-up of a rat among the ruins of a bombed city. Underlying the control of rats was their capacity to destroy and contaminate food, to undermine foundations of buildings and, most significantly, transmit human diseases, including conditions such as murine typhus and plague known from Europe and Asia during the war. The film continued detailing strategies for rat control city block by city block (see Illustration 8.5), advocated hygiene recommendations for responsible citizens and described in detail how to kill rats through poisoned food preparations called 'torpedoes' (see Illustration 8.6). The message of the film oscillated between advising general audiences and addressing hygiene professionals in a teaching film manner. Initially the film was intended for professionals in communities and cities to encourage them to take action, which happened for example in Alberta, Canada three years later. Films were made accessible to wider audiences in schools and associations as well.

The second example *The Good Life (Health Crusade)*, illustrates the second major orientation of health education films of the immediate post-World War II period. Produced by Wessex Films London for the UN Film Board, the motion picture was the last of a series of six films on 'The Changing face of Europe (Great Hope)'. Humphrey Jennings, who was according to Lindsay Anderson 'the only poet in British film', directed the venture. In 1933 Jennings had joined the General Post Office Film Unit created by John Grierson, where he had participated not only in publicity for the British Postal Service but also in developing significant pedagogical intentions and describing the British society and its functioning. Jennings died during the shooting of the film and was replaced by Graham Wallace. The colour film, a technically sophisticated

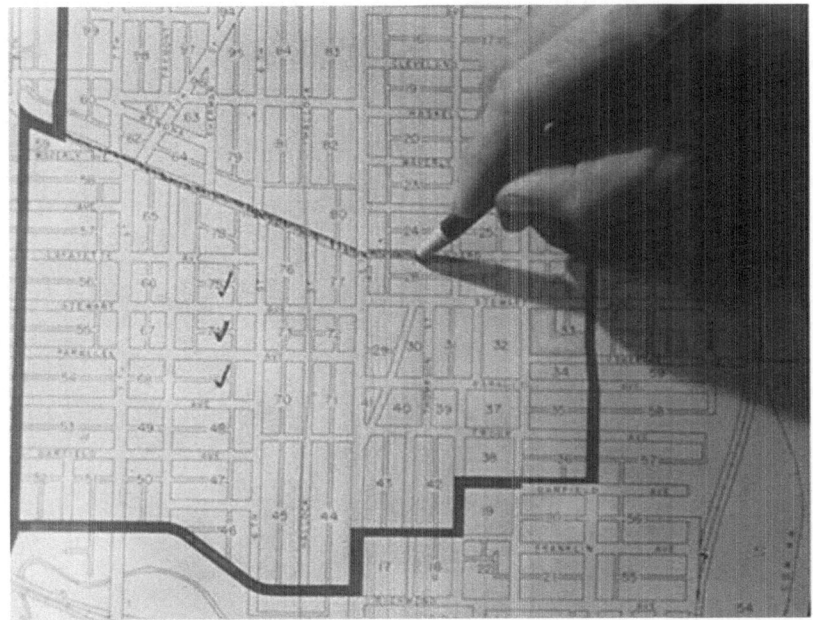

Illustration 8.5 Strategic map of house-block planning for a rat-killing operation. *Practical Rat Control. Rat Killing* (1950). Published courtesy of the National Library of Medicine

production lasting 26 minutes was released in 1951. It opened with the statement that physicians in ancient Greece – referring here directly to Hippocrates and his medical writings – declared that the 'good life' depended on good health. The film, a pet project of Jennings'[39] for which he produced the script and which was initially to be titled simply 'Health', presented health inequalities in post-war Europe and described efforts underway to prevent disease and to promote well-being through international initiatives such as WHO projects[40] or the International Tuberculosis Campaign (ITC).[41] Financed by the Marshall Plan, the film lobbied for international cooperation illustrated in a succession of vignettes: TB vaccination in Greek villages; children in Rome in danger of a host of diseases; in Greece again, at Lamia, the Marshall Plan funds helped build a sanatorium; in Geneva epidemiologists at the WHO tracked disease worldwide; while at Missolonghi the picture screened WHO anti-malaria campaigns employing DDT by airplane and mechanical spraying (see Illustration 8.7) in a long sequence.[42] In order to reach out to remote areas and populations, mobile projection

Illustration 8.6   Preparation of 'torpedoes' mixing food and rat poison as weapon against rat invasions. *Practical Rat Control. Rat Killing* (1950). Published courtesy of the National Library of Medicine

units invented during World War I by the Rockefeller Institution for its mission in Europe were remobilised. Translated into at least five other languages (French, Spanish, Portuguese, Turkish and Greek) the film was aimed at local populations, rural or urban, as part of the UN communication strategies. Health, a seemingly apolitical concern beyond the opposing ideologies of the allied forces, was a genuine humanitarian cause perfectly suited for engagement by the UN and the WHO to reconstruct a democratic and peaceful Europe.

The following two sections will analyse the chosen sanitary propaganda films along two lines. First, the films are increasingly crammed with war metaphors, producing analogies between war combat and fighting diseases. This was not a new way of representing bacteriology from a historical perspective,[43] but film propaganda was re-militarised, rationalised and perfected as an outcome of World War II. The analysis will underline the role and the impact of the extension of war propaganda leading to international health crusades pursuing war in a 'humanitarian'

Illustration 8.7 DDT landscape treatment by airplane in Greece. *The Good Life-Health Crusade* (1951). Published courtesy of the DHVS-IRIST film collection, University of Strasbourg

perspective as a defence against disease vectors. A second line of argument will interrogate the *mise en scène* – the presentation and staging in film of scientific chemical solutions as alternatives to more traditional physical transformations of the environment. From spraying oil on water surfaces as larvae control to DDT 'treatment' of landscapes by plane (see Illustrations 8.1 and 8.7), chemicals were staged as a modern means of transforming environments in order to improve sanitary conditions.

8.3 War on vectors of communicable diseases

Malaria (1931/41) referred only to the 'civilizing work of science and technology in foreign countries and colonies'. War as a theme was absent from the film despite its reediting at the beginning of World War II. The film was centred on one individual disease. If this reflected the corporate strategy behind the film promoting a product in the context of its use – here petrol from Shell – it was at the same time a more general configuration as the history of malaria films shows.

Oil Larviciding started with a scene of oil spraying. The film indicated that oil is 'an effective means of interrupting transmission of mosquito born diseases' by attacking the vulnerable aquatic stages of the mosquito species. Despite its 1943 production date, the film used classical Darwinian language of 'survival of the fittest' in a by then usual framework of struggle over existence between bacteria, parasites and human organisms. War itself as a metaphor or a 1943 reality was notoriously absent from the film. This might seem plausible in a technical teaching film, but as the later *Rat Killing* indicates, that was not necessarily so.

World War II mobilisation triggered heightened concerns for sanitary measures in army camps and led to research for new chemical products in the fight against insects and communicable diseases. Identified as a useful substance in 1939 by Paul Müller for the Swiss pharmaceutical company Geigy, DDT was patented in 1940 and the newly created US Typhus Commission tested its efficacy in early 1943.[44] Quickly DDT became a significant typhus control strategy employed during the last two years of World War II and its application was extended to vector control for other communicable diseases including malaria. The 1947 US War department training film *DDT in the Control of Household Insects* underlined DDT's civilian career. The 19-minute film portrayed how 'DDT was one of the great scientific achievements of the war, and the most powerful weapon in insect control in the household'.[45] Extending environmental sanitation from war barracks to civilian everyday life after war, the film presented educational advice on how DDT was 'a friend and ally' in the fight against mosquito-borne diseases such as dengue, malaria, filariasis and yellow fever.

Films on *Practical Rat Control* around 1950 illustrated the importation of wartime concerns and organisation of sanitary environmental measures into the public civilian sphere. The militarisation of public health films that occurred after the end of World War II integrated and mobilised the war in three specific ways. First, it was only after the end of hostilities that war as a metaphor and its related linguistic military terminology became integrated into technical and educational films that in principal were not concerned with it. In a sense military images and words 'returned home' with the soldiers as they were demobilised establishing continuities between war film propaganda, and health education and information in post-World War II films. Second, it was not just the generic use of a war metaphor but rather World War II as a precise and widely shared existential experience that appeared in health education pictures; war camp organisation and conditions (strategic maps, weapons, rat poisons labelled 'torpedoes', echoing military orders such as 'keep them out') were transferred in both a real and a figurative sense into the civilian sphere of health education propaganda. Films mobilised the victorious outcome of World War II, stressing that after successful defence against human invaders as mentioned in *Rat Killing*, war continued as it was extended to a war on vectors of communicable diseases. War became a device to foster spectator adhesion through the suggested commonly shared wartime experience. *The Good Life* perfectly illustrated this point. Directly building on war propaganda pictures, the film visually presented aircraft power and war mobilisation leading to international health 'crusades' that pursued war from a 'humanitarian' perspective as a defence against disease. Third, staging know-how and products (rat poison, DDT, etc.) as war achievements, films presented them in a certain sense as a return of wartime investments of money and industrial R&D to the civilian society when adapting them for use in the urban environment of industrial cities (*Rat Killing*), or in the everyday household (DDT).

Last but not least in respect to the central issue of this paper, 'defence against disease' took on a very specific meaning in the films portrayed here. 'War on ...' concerned rarely the immediate causal agent of a disease but rather focused on the environmental settings that allowed disease transmission. *The Good Life* perfectly illustrated this point. War metaphors employed 'war on insects and rodents' as potential carriers of microorganisms responsible for disease. Films now particularly targeted intermediary hosts in 'ecological' systems. Consequently films progressively moved from tentative attempts to contain individual diseases (*Malaria*) and precise pathogens, to picturing eradication of animal species that were considered to be environmental living factors for multiple 'communicable diseases' and participating in complex disease transmitting life-cycles. Environment was conceived here rather as a living space for undesirable vectors than as a genuine resource and condition for human existence.

8.4 Chemical treatment of the environment

A second line of argument emerging from a comparative study of the films concerns the question of *mise en scène* of scientific chemical solutions 'treating environment'. The third section of the film *Malaria* staged most clearly a double shift of approach: from treating humans to treating the environment and from physical landscaping to chemical treatment. Indeed, the film displayed first scenes of physical engineering of landscapes (drying wetlands, drainage, etc.) employing considerable groups of workers. Then it moved to presenting 'modern' petrochemical solutions depicted by a sole individual spraying (see Illustration 8.1). From groups of cowed colonial workers to an individual spraying operator standing immobile and straight up, the images implied that the job was being done by the substance and not the operator anymore. The film contrasted the reduction of workers necessary to 'treat the environment' (from many to one) with hygiene being transformed from private to public. Images were shifting from medical prevention of individuals taking quinine[46] as pharmaceutical prophylaxis[47] to a public health approach where treating the environment equalled improving health conditions for all members of a population. Chemical treatment of the environment was thus depicted as the transition from physical landscaping and medical treatment of individuals to environmental approaches based on chemistry. The chemistry at hand was situated somewhere between pharmaceuticals and agricultural insecticides. It is noteworthy that it was precisely the Shell Company (sponsoring the film *Malaria*) that was engaged in insecticide production at Colorado's Rocky Mountain Arsenal, the same site where the US army produced chemical weapons during World War II.[48] *Malaria* concluded that the aim of this 'chemical modernity based on the scientific laboratory' (see Illustration 8.2) was the total eradication of pathogen intermediary host populations, meaning here the mosquito.

The second part of the PHS training film *Oil Larviciding* described technical solutions and scaling-up of spraying – designated as power oiling with trucks and boats (see Illustration 8.4). Typical of the overall evolution of similar films was their pragmatic technology – and chemistry-based problem-solving approach – be it oil larviciding, rat killing or DDT spraying. *Oil Larviciding* eventually moved from reflecting late 1940's work hazard concerns to the relative three disadvantages of oil: '(1) it was heavy and hard to transport; (2) it often saturated the clothes of labourers, irritating their skin; and (3) it formed an objectionable scum at the water edges'. Material organisational difficulties, workers' health problems and environmental concerns (if they can be interpreted that way) figured here without hierarchical distinction but of equal importance. Occupational health and environmental concerns were mentioned in a common set of concerns pointing to the now classical link between the two fields.[49] This more critical and reflexive perspective on oil appeared at a time when the already industrially produced DDT's career as a replacement of oil was dawning.

The chronologically last film of our corpus brings us back to Europe and the progressive organisation of the 'United Nations of Health'. In

July 1948, the year of the creation of the WHO, the UN Film Board produced a first feature length film presenting the general philosophy, organisation and mission of the WHO: *The Eternal Fight* (1948). Under the initial dictum 'Deliver us ... from contagious diseases' the film portrayed what WHO officials designated as 'the new global threat of epidemics' in the age of industry, mass transportation and mass communication. Two-thirds of the way through the *The Good Life-Health Crusade* (1951), the camera swings to Missolonghi 'a small village on the shore of the Ionic sea that had a very difficult time keeping its head out of the putrid waters surrounding it'. If physical landscaping and drainage were evoked in off-voice commentary, the corresponding images focused on the DDT treatment by plane of the entire area (see Illustration 8.7). The commentator reassuringly asserted: 'and when DDT will have completed its job on this land haunted since ancient times by misery and death, people will cultivate rice and life will flourish again'. As in *Malaria* it is the personified substance that accomplished the job. Bridging past and present, east and west, north and south the humanitarian hymn concluded with the ancient Greek precept 'a healthy soul in a healthy body' as the camera glorified architectural achievements of reconstructed Europe through bridges, social housing blocks and modern power plants. This vision of modernity was shared by both *Rat Killing* and *The Good Life* (see Illustration 8.8) in their pictorial conclusions.

From spraying petrol on water surfaces as larvae control to DDT 'treatment' of landscapes by plane, chemicals were staged as modern means to fight communicable diseases. In health propaganda films after World War II such as *Rat Killing* and *The Good Life* they were part of the immense effort to reconstruct Europe and the US both morally and physically after the war. Chemicals and especially DDT were icons of chemical modernity comparable to bridge constructions signifying infrastructure or modern buildings symbolising social housing. Here again World War II served to model perspectives that would be decisive for two decades to come. Not only bombshell power but also chemical ingenuity was celebrated as the foundation for the reconstruction of 'a better world', defeating first Nazi Germany then infectious diseases. At the centre lay the firm conviction that the chemical treatment of the environment was the most promising approach to the problem. A conviction further illustrated by the WHO Sardinian Project for total mosquito eradication by DDT on the island and propagated by film media[50] until the criticism voiced by counter-cultures in the 1960s.[51]

Illustration 8.8 Final image of the film on rat control symbolising the opposition of traditional and modern housing solutions in urban environments. *Practical Rat Control. Rat Killing* (1950). Published courtesy of the National Library of Medicine

8.5 Conclusion

The present paper has addressed the question of how public health interventions in rural and urban environments were presented to the public in general health education and training films in the US and Europe between 1930 and 1960. Two major themes crystallise. First, films mobilised metaphors, especially after World War II, and extended the war efforts to insect and rodent eradication. The second theme established a long tradition of film propaganda – be it corporate, state or international agency-sponsored deploying a *mise en scène* advocating scientific chemical solutions to 'treat the environment'. Productwise, historical scholarship has so far singled out DDT and its post-war hegemony but films underline continuities with a longstanding tradition from oil to DDT. In contrast to the aftermath of World War I which triggered criticism of techno-scientific modernity, post-World War II films celebrated chemistry as a powerful strategy to construct a 'better world' liberated both from Fascism and major lethal diseases. Public health environmental interventions targeted especially animal species identified as host organisms and disease reservoirs moving away from the turn of the century reductionist pathogen-disease paradigm of the golden Pasteur-Koch age of bacteriology. Alternatives to more traditional physical engineering of the environment, chemical solutions combined and adapted approaches of pharmaceutical chemotherapy as individual prophylaxis with agricultural practices using insecticides for parasite eradication. More generally speaking, this *mise en scène* was part of the aesthetic, technical and educational aspects of what has been described as 'the post-war fascination with, and faith in, technology and the ability to transform the world'.[52] Health education films themselves were as modern and sophisticated communication technologies a war-derived civilian product, an offspring of wartime propaganda. Their use beyond a simplistic understanding as illustration of agency's work highlights their modernist technological complicity with what they promoted and their tacit agreement to serve international science expert hegemony in their role as mediators between professional institutions and the public.

Filmography

Malaria, produced by Shell Corporation, approximately 1931/41. Directed by Arthur Grahame and Ethan Tharp, microcinematography by Percy Smith, Black and White, 20 minutes. Probably a re-edition of *Malaria* produced by the Shell Film Unit in 1931.

Oil Larviciding, Series Training films, produced by the US Public Health Service, 1943. Director anonymous, colour, 12 minutes.

Practical Rat Control. Rat Killing, produced by the US War Office, with the advice and assistance of the Communicable Disease Center, the USPHS, and the Federal Security Agency, 1950. Director anonymous, Black and White, 12 minutes.

The Good Life (Health Crusade), Series The Changing Face of Europe (Great Hope), produced by Wessex Films London, 1951. Directed by Humphrey Jennings and Graham Wallace, Colour, 26 minutes. The screened French version of the film carries the title *Croisade pour la santé*.

Notes

1. A. Cunningham (1992). 'Transforming plague. the laboratory and the identity of infectious disease', in A. Cunningham and P. Williams (eds). *The Laboratory Revolution in Medicine*, (Cambridge: Cambridge University Press), pp. 209–24; G. Geison (1995). *The Private Science of Louis Pasteur*, (Princeton: Princeton University Press); A. Mendelsohn (1996). *Cultures of Bacteriology: Formation and Transformation of a Science in France and Germany, 1870–1914* (Princeton

University: PhD Thesis); C. Gradmann (2009). *Laboratory Disease: Robert Koch's Medical Bacteriology* (Baltimore: Johns Hopkins University Press).

2. J. J. Duffy (1990). *The Sanitarians. A History of American Public Health*, (Urbana: University of Illinois Press); V. Berridge (1999). *Health and Society in Britain since 1939*, (Cambridge: Cambridge University Press); B. Latour (1988). *The Pasteurization of France* (Cambridge: Harvard University Press).

3. L. Slater (2009). *War and Disease. Biomedical Research on Malaria in the Twentieth Century* (New Brunswick, New Jersey and London: Rutgers University Press).

4. A. Mendelsohn (1998). 'From eradication to equilibrium: how epidemics became complex after World War I', in C. Lawrence and G. Weisz (eds). *Greater than the Parts: Holism in Biomedicine, 1920–1950* (Oxford: Oxford University Press), pp. 303–31; E. Tognotti (2003) 'Scientific triumphalism and learning from facts. bacteriology and the "Spanish flu" challenge of 1918', *Social History of Medicine*, 16, 97–110.

5. Leo Slater considers that the divide between pursuit of the disease organism versus its vector has separated biomedical and environmental and approaches to malaria. Evidently this definition reflects Slater's central interest in the US anti-malarial program (1939–46) focussed on the parasite and chemotherapy rather than on ecological conceptions of disease associating malaria with wetlands, mosquitoes, drainage and DDT. Oiling is surprisingly absent from Slater's account. L. Slater (2009). *War and Disease*.

6. For linguistic clarification it needs to be stressed here that during the interwar period 'propaganda' had a positive connotation as promotional technique for mass information and instruction.

7. S. Lederer and N. Rogers (2000). 'Media', in R. Cooter and J. Pickstone (eds). *Medicine in the Twentieth Century* (London: Harwood), pp. 487–502.

8. L. Reagan, N. Tomes and P. Treichler (2007). *Medicine's Moving Pictures. Medicine, Health, and Bodies in American Film and Television* (Rochester: Rochester University Press).

9. For a criticism of documentary film as mere presentation of a scientific and technical "reality" see: C. Delage and V. Guigueno (2004). *L'historien et le film* (Paris: Gallimard); U. Jung and M. Loiperdinger (2005). *Geschichte des dokumentarischen Films in Deutschland. Band 1 Kaiserreich 1895–1918* (Stuttgart: Reclam); T. Boon (1999) *Films and the Contestation of Public Health in Interwar Britain* (London: PhD Thesis University of London).

10. C. Bonah and A. Laukötter (2009). 'Moving pictures and medicine in the first half of the 20th century: Some notes on international historical developments and the potential of medical film research', *Gesnerus*, 66, pp. 121–45.

11. T. Boon (2008). *Films of Fact: A History of Science in Documentary Films and Television* (London and New York: Wallflower Press).

12. A. Nichtenhauser adopts a similar structure for his 1954 manuscript on the History of Medical Film using subsections for different countries. The Nichtenhauser manuscript is deposited at the National Library of Medicine, Bethesda. A. Nichtenhauser (1954). 'A History of Medical Film', unpublished manuscript, National Library of Medicine, Nichtenhauser papers, Washington DC.

13. R. M. Barsam (1973). *Nonfiction Film. A Critical History* (New York: E.P. Dutton), pp. 206–7.

14. J. C. Ellis and B. A. McLane (2005). *A New History of Documentary Film* (New York: Continuum), pp. 208–26.
15. L. Cartwright (1995). *Screening the Body: Tracing Medicine Visual Culture* (Minneapolis: University of Minnesota Press).
16. U. Jung and M. Loiperdinger (2005). *Geschichte des dokumentarischen Films in Deutschland. Band 1 Kaiserreich 1895–1918* (stuttgart/Ditzingen: Reclam).
17. T. Lefebvre (1993). 'Contribution à l'histoire de la microcinématographie: de François-Franck à Comandon', *1895*, 14, pp. 35–46; I. O'Gomes (1994). 'L'œuvre de Jean Comandon' in A. Martinet (ed.). *Le cinéma et la science* (Paris: CNRS éditions), pp. 78–85.
18. M. S. Pernick (1978). 'Thomas Edison's tuberculosis films: mass media and health propaganda', *Hasting Center Report*, 8, pp. 21–7.
19. U. Keitz (2005). 'Wissen als Film. Zur Entwicklung des Lehr-und Unterrichtsfilms' in K. Kreimeier, A. Ehmann and J. Goergen (eds). *Geschichte des dokumentarischen Films in Deutschland. Band 2 Weimarer Republik 1918–1933* (Stuttgart: Reclam), pp. 120–50.
20. J. C. Ellis and B. A. McLane (2005). *A New History of Documentary Film*, pp. 57–63; R. M. Barsam (1973). *Nonfiction Film*, pp. 37–80; R. Low (1979). *The History of British Film 1929–1939: Documentary and Educational Films of the 1930s* (London: George Allen and Unwin).
21. J. Goergen (2005). 'Industrie und Werbefilme', in K. Kreimeier, A. Ehmann and J. Goergen (eds). *Geschichte des dokumentarischen Films*, pp. 33–8; J. Goergen (2005). 'In filmo veritas! Inhaltlich vollkommen wahr. Werbefilme und ihre Produzenten', in K. Kreimeier, A. Ehmann and J. Goergen (eds). *Geschichte des dokumentarischen Films*, pp. 348–63.
22. R. M. Packard (2007). *The Making of a Tropical Disease: A Short History of Malaria* (Baltimore: Johns Hopkins University Press); L. Slater (2009) *War and Disease*; M. Humphreys (2001). *Malaria in the United States: Poverty, Race and Public Health* (Baltimore: Johns Hopkins University Press); F. M. Snowden (2006). *The Conquest of Malaria: Italy, 1900–1962* (New Haven: Yale University Press, 2006); M. Cueto (2007). *Cold War, Deadly Fevers. Malaria Eradication in Mexico, 1955–1975* (Washington: Woodrow Wilson Center Press/Baltimore: Johns Hopkins University Press).
23. M. Fedunkiw (2003). 'Malaria films: motion pictures as public health tool', *American Journal of Public Health*, 93 (7), pp. 1046–57. In a historical analysis of malaria films exclusively drawing upon US and British productions Fedunkiw distinguishes three periods: 1912–30 with 16 identified films produced being the period of early instruction and prevention; 1930–62 with 139 films mainly dedicated to fighting against the disease and vector eradication; and eventually 1962–82 with only 15 films, indicating a certain disillusionment with educational propaganda. Fedunkiw's analysis is based on: List of public health films, Rockefeller Archive Centre, RF RG1 A83, 100, Films 1941–1945, Box R 1996. List of films on various public health topics. VI Mosquitoes, Malaria & Malaria Control, 16–24. At the present a copy of the Rockefeller film has not been located.
24. The first film is *Malaria*, co-produced by UFA and the company Bayer, directed by Ulrich Kayser, script Karl Spielmann, camera Erich Menzel, animation Bernhard Huth and music by Walter Schütze. Black and White, sound, 1934, 24 minutes. A second film, *Malaria. Experimentelle Forschung und*

klinische Ergebnisse, is presented as produced by the scientific-pharmaceutical Department of the Bayer Corporation, Leverkusen. Black and White, sound, no date (approximately 1934). It seems to be the chemical compound-oriented complement to *Malaria* (UFA/Bayer) which is disease and public health-oriented.

25. O. Wagner (1934). 'The cinematograph in the service of medical and biological research in medicine in its chemical aspects', *Reports from the Medico-Chemical Research Laboratories of the IG Farbenindustrie Aktiengesellschaft*, pp. 391–404.

26. S. Legg (1954). 'Shell film unit: twenty-one-years', *Sight & Sound*, 23, pp. 209–11.

27. The film is probably a sound re-edition of the first malaria film produced by the Shell Film Unit in 1931. The original 1931 edition is considered as lost. M. Fedunkiw has identified a second version with identical title dated in the Rockefeller list to 1939 and distributed by the New York Office of the Dutch Shell Oil Company but the film description makes it clear that it differs from the 1931/1941 version. A copy of the 1931/1941 film preserved in the Historical Medical Film Collection at the University of Strasbourg was screened and serves as the basis for the following analysis.

28. The 1944 film *You too can get Malaria* produced by Verity, part of the British Directorate of Army Kinematography even used insufficient human compliance as a plot device for a malaria public health education film telling the fictive story of a new army recruit willingly resisting individual preventive measures and consequently falling ill of the disease. M. Fedunkiw (2003). 'Malaria films'.

29. Most accounts on malaria underscore the place of oil larviciding in the 1930s and jump directly from the chemical Paris green (copper acetoarsenite) developed in 1921 in the US by M. A. Barber, T. B. Hayne and W. Komp to the use of DDT during World War II. L. Slater (2009). *War and Disease*.

30. *Malaria* (Shell), 18th minute.

31. *Malaria* (Shell), 20th minute.

32. M. Barsam (1973). *Nonfiction Film*, p. 13.

33. *Oil Larviciding*, 1943. The film is preserved at the National Library of Medicine (NLM) at Bethesda/Washington DC. NLM ID: 8800314A. I would like to thank David Cantor and Stephen Greenberg for their archival assistance and kind help at the NLM.

34. *Defense against Invasion* produced by Disney Corporation, 1943. Distributed after 1945 by the US Information Service.

35. Anonymous (1947) 'UNESCO's World Plans', *Documentary News Letter*, 6/55, 67.

36. The film is preserved at the NLM ID: 9423674.

37. *Keep them out* is produced by Stark-Films in cooperation with the USPHS, 1942. Black and White, sound, 10 minutes. The film is preserved at the NLM. ID: 8700204A.

38. Other titles in the series produced the same year are: *Rat Ectoparasite Control* (1950). NLM ID: 9422793. *Sanitation Techniques in Rat Control* (1950). NLM ID: 9301354A. *The Rat Problem* (1950). NLM ID: 9421645. *Habits and Characteristics of the Rat: The Roof Rat* (1950). NLM ID: 9423318. *Practical Rat Control: Ratproofing* (1950). NLM ID: 9423676. All of these films are produced in 1950 by the US War Office/Army, with the advice and assistance of the Communicable Disease Centre, the USPHS, and the Federal Security Agency.

39. Films by Jennings include *The Story of a Wheel* and *Spare Time* on workers leisure time. During World War II Jennings directed *The Heart of Britain*, a documentary on London being bombed. http://www.Marshallfilms.org, date accessed 20 February 2010. R. M. Barsam (1973). *Nonfiction Film*.

40. S. Lee (1997). 'WHO and the Developing World: The Contest for Ideology', in A. Cunningham and B. Andrews (eds). *Western Medicine as Contested Knowledge* (Manchester: Manchester University Press), pp. 24–45.

41. N. Brimnes (2007). 'Vikings against tuberculosis: the international tuberculosis campaign in India, 1948–1951', *Bulletin of the History of Medicine*, 81, pp. 407–30. G. W. Comstock (1994). 'The international tuberculosis campaign: a pioneering venture in mass vaccination and research', *Clinical Infectious Diseases*, 19 (3), pp. 528–40.

42. D. H. Stapleton (1998). 'The dawn of DDT and its experimental use by the Rockefeller Foundation in Mexico, 1942–1952', *Parassitologia*, 40, pp. 149–58. D. H. Stapleton (2005). 'A lost chapter in the early history of DDT: the development of anti-typhus technologies by the Rockefeller Foundation's Louse Laboratory, 1942–1944', *Technology and Culture*, 46, pp. 513–40; S. Bertsch McGrayne (2001). *Prometheans in the Lab. Chemistry and the Making of the Modern World* (New York: McGraw Hill); P. Brown (1998). 'Failure as success: multiple meanings of eradication in the Rockefeller Foundation Sardinia Project, 1946–1951', *Parassitologia*, 40, pp. 117–30; M. Cueto (2007). *Cold War, Deadly Fevers*.

43. P. Sarasin (2003). 'Infizierte Körper, kontaminierte Sprachen. Metaphern als Gegenstand der Wissenschaftsgeschichte', in P. Sarasin (ed.). *Geschichtswissenschaft und Diskursanalyse* (Frankfurt am Main: Suhrkamp), pp. 191–230. C. Gradmann (2007). 'Unsichtbare Feinde. Bakteriologie und politische Sprache im deutschen Kaiserreich', in P. Sarasin et al. (eds). *Bakteriologie und Moderne. Studien zur Biopolitik des Unsichtbaren*, (Frankfurt a.M: Suhrkamp), pp. 327–54.

44. D. H. Stapleton (2005). 'A lost chapter in the early history of DDT'.

45. *DDT in the Control of Household Insects* (1947). Produced by the US War Department. Colour, 19 minutes.

46. *Malaria* (Shell), 16th minute.

47. L. Slater (2009) *War and Disease*, pp. 17–38.

48. E. P. Russell (1999). 'The strange career of DDT: experts, federal capacity, and environmentalism in World War II', *Technology and Culture*, 40.4, pp. 770–96. For links with the Shell Film Unit see: E. Barnouw (1993). *Documentary. A History of the Non-Fiction Film* (New York: Oxford University Press), pp. 213–15.

49. C. Sellers (1997). *Hazards of the Job: From Industrial Disease to Environmental Health Science* (Chapel Hill, University of North Carolina Press).

50. *The Sardinian Project* (1949). Produced by Shell, Nucleus Film Unit & Erlass. 36 minutes.

51. E. P. Russell, (2001). *War and Nature. Fighting Humans and Insects with Chemicals from World War I to Silent Spring* (New York: Cambridge University Press).

52. R. M. Packard (1998). '"No other Legal Choice": Global Malaria Eradication and the Politics of International Health in the Post-war Era', *Parassitologia*, 40, pp. 217–29, here p. 220.

9

Cross-Nationalising the History of Industrial Hazard

Christopher C. Sellers

The recent furore over lead-contaminated toys points to how transnational our encounters with industrial hazards have become. Over the summer months of 2007, Mattel, Inc., the world's largest toymaker, recalled some 20,000,000 toys, nearly 3,000,000 of them because of lead-contaminated paint. The recall mushroomed into an international event. In Germany regulators pulled some 1,000,000 toys from the shelves, in Britain and Ireland, 2,000,000; countries from Malaysia to Bahrain joined the toy returns. In just one of the incriminated Chinese factories, some 83 different kinds of toys may have been painted with lead pigment. Suddenly, this toxic metal, along with carbon monoxide, the world's oldest recognised industrial hazard, took the Western world by surprise. Despite widespread assumptions that we were safe from its clutches, a whole new vein of lead had turned up, running into our department stores, homes and perhaps also our children. The US coverage was split between blaming and exonerating Mattel's executives, but more uniformly, it construed this as a crisis for American consumers. Coverage did extend to the industrial pollution produced by China's quarter-century of economic boom. But the insides of its factories were another matter, largely neglected.[1]

To help us better understand the past out of which such an episode has sprung, this paper argues for the need to cross-nationalise our histories of industrial hazards. Unpacking our modern ways of looking at the lead hazard and contrasting these with how we used to see it, I assert that we need approaches that are not just comparative but that also capture the changing ways those in differing nations interacted with one another. In particular we need studies that can enhance our understandings of the relationship between hazard histories in the developed versus the developing world.

A century and a half ago, when Karl Marx coined the notion of commodity fetishism, he was referring to a phenomenon that happened inside 'societies in which the capitalist mode of production prevails'. For many of us today, as well, the spell of the commodity itself, shiny and colourful in its appearance on store shelves, obscures the social relations of the workplace, as Marx would have it, including labouring hands that actually made it.[2] How much more powerful the spell of the commodity has become in our own time and place, when most manufacturing happens in a foreign land, when American and European encounters with the factory and its work are becoming so few and far between. We have heard much about how globalisation has collapsed distance, linking people from different nations via, among other ways, global supply chains. But you do not need to be a Marxist to see how the cross-national linkage of the supply chain can enable attitudes not so much of connection as of indifference. Undertaking the task of making our things, these distant places shoulder a scale and variety of hazards that our own 'progress' has made it ever more difficult for us to imagine.

Health and other historians, for all the work they have done over the past 20 years in uncovering the history of these hazards, have thus far provided relatively few insights into this contemporary dilemma. This paper explores ways we may give greater heed to those cross-national and mutual transformations that have brought us to this juncture. Study of variations and changes beyond the bounds of any single nation state, in the material flow, perception and regulation of toxins may profit from recent methodological innovations in business, labour, technology, environmental and urban history. I suggest two investigatory strategies as especially useful. First, we may explore more multi-scalar modes of analysis. Ranging between the local, regional, national and global will help to spotlight historical relationships *between* places, however contrasting and divergent their trajectories. Second, we may focus especially on toxins' materiality. Such an emphasis can bolster our exploration of differences in their perception across time and place, I suggest, especially when paired with a principle of 'environmental symmetry', rooted in the understandings of today's science. As a way into these possibilities, the paper then sketches some little-explored approaches to the twentieth-century history of the lead industry and its regulation, in particular, through the 'local' lens of a lead smelter along the border of the United States and Mexico.

9.1   The multi-scalar and the material: historiographic precedents and lead's present

Nowhere have the contrasts between more global or transnational and more national or local approaches been more marked than in the historiography of industrial hazards itself. When Henry Sigerist and Ludwig Teleky wrote about workplace hazards in the middle of the twentieth century, their analyses leaped with relative ease from one nation's experience to another's. Their inclusion of so many different nations' experiences side by side hinged on largely unscrutinised assumptions about the cross-national uniformity of industrialisation, its hazards, and their prevention. Also ignoring extra-workplace hazards, they homed in especially on two aspects of workplace hazards: their detection and regulation. Both conceptualised this 'progress' mainly as the passage of the right laws, creating rules and agencies that empowered experts, who accomplished what was needed.[3] The revival of historiography of industrial hazards from the 1980s greatly complicated this earlier intellectual legacy of nation hopping. The methodology of social history revealed just how many other groups have had an influential role in determining the approach to occupational hazards in a given place and time. Beyond doctors and hygiene experts, we have learned, the willingness of managers or owners, technological change, and especially the insistence or protest of workers, whether unionised or not, may all impact on the dangers afflicting a given time and workplace. Within any given nation or industry, questions of recognition and regulation are so fraught with struggle between groups that 'progress' itself looks far more contingent, a matter of contention.[4]

Developments in other fields, as well, have complicated any aspirations to situate the industrial hazard history of different nations alongside one another. Environmental history, for one, has broached those many other hazards and conflicts that industrial production may impose beyond the confines of workplaces. Factories may also contaminate nearby neighbourhoods, or via air or rivers, more distant locales. Still further away, their toxins may also turn up in toxic consumer goods, from paint to apples.[5] Moreover, tools for supranational history that have been around for a while – the tradition of comparative case studies, world-systems theory, 'global' history and the history of international institutions – need some tweaking to be adapted to the largely more localised insights of the last decades of social and cultural history. Often, though not always, choices of scale have inclined scholars toward one side or another of the nature-culture divide. Those traditions of more sweeping and macro historical visions have often leaned on natural or material dynamics, whether of infection or economics. The more micro focus of social and cultural history of the last couple of decades of scholarship has meanwhile yielded other kinds of insights, into those many groups, perceptions and conflicts shaping the history of health and environment in a single place.

Complicating matters, too, has been the cultural turn of scholarship of the 1980's and 1990's history of science as well as medicine. Scholars in science studies laid early groundwork, by urging then largely by assuming an agnosticism toward the natural objects named and debated by scientists, early on styled as a sociological 'principle of symmetry'. To get at the social and power dynamics of scientific disputes, scholars were to treat both sides as having equally compelling claims about the natural world, without regard to who won or lost. Across science studies as well as the history of science and medicine, much cutting-edge work of the 1980s and 1990s went on to shun invocations of the material, whether natural or economical, as Popperian positivism. Whether more or less influenced by the ensuing 'science wars', historians of science and medicine often subscribed to a rigorous historicism with similar implications: not to attend to material objects and influences until the doctors and scientists of an era recognised them.[6]

Partly to get beyond the stark position-taking of these debates, scholarship in sciences studies as well as medical and science history has more recently gravitated toward questions of ontology, that branch of philosophy that deals with the experience and perception of objects in our world. Recognising that more than just 'ideas' are at stake in the historical trajectories of medical and other sciences, scholars of technoscience now prefer to talk of 'materiality': how differently bodies – human or otherwise – have manifested themselves, or mattered, across different places and times. Projects such as Lorraine Daston's edited volume on biographies of scientific objects and Ian Hacking's for historical ontology seek to reconcile the dynamic historicity of scientific objects with their solidity and durability for historical actors in a given time and place. This is powerful insight, of great value to the historian of industrial hazard. Yet such an historian would hobble him or herself considerably by sticking, as Daston and Hacking do, to 'objects or their effects which do not exist in any recognisable form until they are objects of scientific study'.[7] A troubling but enduring constant of industrial hazard history is that the hazards themselves, along with their victims, have often existed long before they were made the objects of formal scientific study. To make the most of an historical ontology approach, historians of industrial hazard would best seek ways of integrating its insights with the health and ecological as well as economic dynamics of the times, facets of history which today's sciences themselves may illuminate.

Easing the prospects for integration, over the last 20 years, veins of enquiry friendlier with 'positivist' sciences have forged new understandings of the history of cross-national economic interchanges, and also, more socio-cultural approaches. Business and economic historians tackling "globalisation" have been joined by those working in world systems theory pioneered by Immanuel Wallerstein, as well as labour and environmental historians, sociologists, and geographers, in seeking to explain the changing global distribution of particular industries over the nineteenth and twentieth centuries, and their social and environmental effects.[8] Especially useful for historians tackling the changing geography of industrial hazards are notions of a 'spatial fix', or more specifically, a 'race to the bottom'. Once technologies of production become portable, producers seek out the cheapest and least protected or resistance labour and locales.[9] Labour historians like Jefferson Cowie have offered compelling models of how to integrate a transnational scope with the localism of social history, though with little to say about any resulting hazards. A few environmental historians have also now framed studies that are similarly transnational, more about the interchange between select nations. Focussing on single commodities, environmental historians such as John Soluri on bananas and Jennifer Anderson on mahogany, have honed ways of studying those cross-national flows of funds, corporations and technology, with a view to the more localised transformations that have accompanied them.[10] As some environmental historians have absorbed insights from cultural history and the history of science and medicine, they have also established more common ground with a "materiality" agenda also being developed by historians of human and public health.[11]

In the history of industrial hazard, much further investigation is needed of just how differently toxins and their effects have been viewed across times and places other than our own; yet strict nominalism and historicism can only take us so far into these differences. In particular, they cannot tell us much about where and when people may remain un- or only partly aware of actual ongoing effects and pathology, or else apprehended these, but in ways radically different from our own. By the same token, of course, a strictly ecological or epidemiological approach tells us little about such crucial issues as the evolution of hazard awareness and understanding, or the sensory and cognitive experience of pollution, or the power struggles that could result. My point is this: any approach that remains too exclusively socio-culturalist *or* materialist will wind up marginalising burning questions in the history of an industrial toxin such as lead. I suggest we may better reconcile what have heretofore seemed diametrically opposing terms of historical analysis by considering a toxin's variable and evolving materiality. The 'stuff' or nature of lead is, after all, what links today's Chinese lead workers with the leaded toys discovered on our store shelves. Differences in culture or perception are vital here, yet now, as in the past, real but less recognised damage may also be at stake, in places that can be quite distant from our own. Better to acknowledge and address such differences and their moral implications; we cannot confine ourselves to social or cultural differences alone. We also need to embrace, more or less forthrightly, what today's scientists now know about lead's impacts on human bodies and environments.

Through a more self-conscious juxtapositioning of lead's contemporary with its historical manifestations, we open the door to a more robustly interdisciplinary holism, a fuller integration of natural scientific and medical insights with those from the social sciences and humanities. We also gain new analytical leverage for juxtaposing and interpreting that great variety of visions brought to bear on a toxin such as lead across time and place. More fully to fathom the variety of ways in which an industrial hazard has 'mattered' across place and time, some privileging of today's scientific consensus about the nature of lead and its physiological effects may well be necessary. Projecting that knowledge across place and time helps us decipher just which societies and groups have grappled with lead, and where we might expect its effects to register with scientists and doctors, but also lay people. In particular, such extrapolations may provide important platforms for extending the historical study of industrial hazards to the developing world. Especially across a divide as deep and complex as that between developed and developing nations, it is important to delineate differences not just in political economy or culture but in concrete contacts with lead and its effects.

As a starting point, rather than a symmetric agnosticism toward scientific claims about exposure, I suggest we include in our historical studies a self-conscious effort to assess historical exposures and their differences: a principle, instead, of *environmental symmetry*. Situating actual exposures to lead today, as experienced and understood in today's United States or Great Britain, over against those in other places and times, can offer vital groundwork for fathoming ways of apprehending this toxin and its effects that may also be quite different from our own.

A principle of environmental symmetry can aid in understanding our own supra-national present, as well as the past. Seen in terms of the longer natural history of our species, human manipulations of the metal over the last two or three hundred years have spurred a lengthy episode of exposures that are much higher than those throughout humans' longer ecological and evolutionary history.[12] Nevertheless, in taking such a sweeping view, such perspectives flatten out that very cross-national diversity that those toxic toys reveal at the heart of today's environmental dilemmas. They thereby offer only limited analytic purchase on changing cross-national patterns, not just of lead exposure but of its control, and the political, economic and cultural differences that make this control possible. By alternating between scales, from the global to the national all the way down to local factories and their surroundings, we can better flesh out those parallel, evolving fusions of culture and nature that comprise lead's twenty-first century present, as well as its twentieth-century history.

We may, for instance, assume an ecological or environmental symmetry to the lead now suddenly and surprisingly headed our way on Chinese toys. It suggests that this lead paint has already wrought tolls elsewhere, not just on the muscles and brains of China's children but of its workers. Such hazards are part of the reason International Labor Organisation (ILO) and World Health Organisation (WHO) officials estimated a global burden of some 1,100,000 deaths from work-related disease and injury alone in 1999, roughly the same toll as from malaria. They project these numbers will rise over the first half of the twenty-first century.[13] A chief reason cited for this future projection is a continued 'race to the bottom'. Corporations, so the story goes, have been hunting not just for the cheapest labour, but for hazard havens, where dangerous materials and processes are more tolerated.[14] One advantage of following this process via lead itself, as a kind of tracer, is that it allows us to consider impacts that have gone largely noticed.

For instance, America's public health triumph over lead is often illustrated through graphs that show declining blood lead levels over the late twentieth century, especially after lead was banned in gasoline.[15] But if we step upward in scale, to look at world lead production itself (see Figure 9.1), a quite different picture emerges. Over the twentieth century, even after the recognition and regulation of the 1970's and 80's, the amount of lead produced has, at best, only plateaued. That recent episode of lead-contaminated toys makes it easy to predict what might be happening if we break this down by nation: Mexico's production rises some, but China's sky-rockets (see Figure 9.2), especially over the last quarter century. What is surprising, though, is that lead production has also not declined in many of the most developed Western nations such as the United States and the United Kingdom (see Figure 9.2). Much more of it comes through secondary production, that is, through recycling of lead scrap. But the trend has still been level or upward.

Why? To explain, we first need to draw some distinctions between realms and varieties of lead exposures involved. The chart of falling lead levels only shows what lead, of all that which Americans continued to surround themselves with, registered in their blood streams. Its fall reflects, more than anything else, a ban on tetra-ethyl lead gasoline, that most efficient means of making the lead Americans bought and consumed 'bio-available', that is, readily absorbed into their bodies. Recognising that the relatively tiny amounts of lead spewed into the atmosphere as car exhaust yielded such high bodily burdens, the US moved quickly from the 1970s onward to ban this as well as other consumer uses that made this toxin such a physiological, as well as ecological danger. But while such measures, along with workplace reductions, brought down blood levels of lead in the United States, they did little to prevent American usages of this toxin that did not make it bio-available,

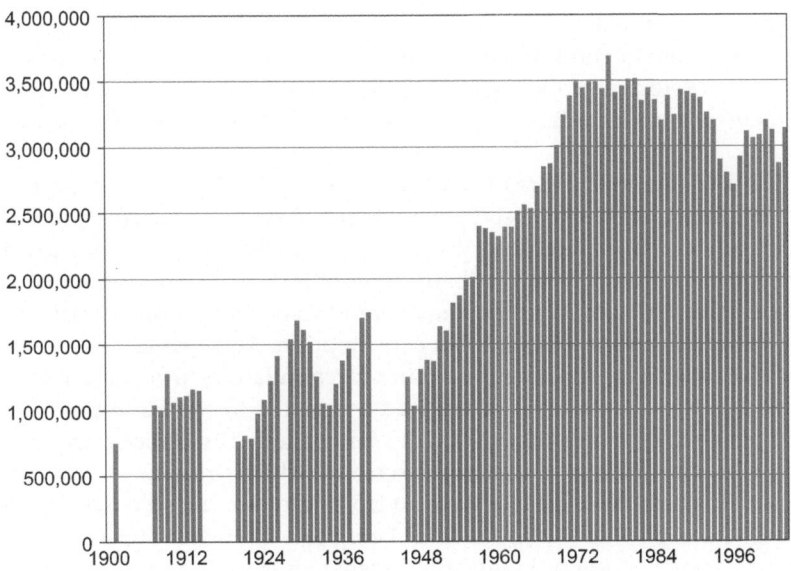

Figure 9.1 Lead production trends worldwide. International Historical Statistics (2007)

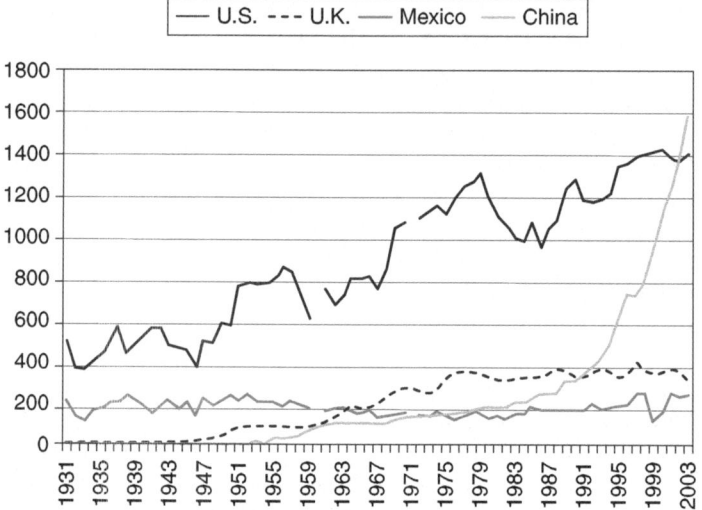

Figure 9.2   Lead production trends by nation. International Historical Statistics (2007)

either in workplaces or consumer products. And they did still less to protect those beyond America's shores, either in the consumer markets of the developing world, or in its mines, smelters and production lines, where the commodities purchased by Americans were increasingly made.

As a result, consumers in the US and other developed-world nations have surrounded themselves with what may be a *growing* amount of lead. But it is 'safe' lead, in products where it is tightly enclosed. Between 1973 and 1993, dissipative uses of lead, from gasoline additives to ammunition to paint and similar products, shrank from 20 per cent to 5 per cent of Americans' lead consumption. Meanwhile, our electric soldering and especially our batteries (accumulators) also contain lead, and batteries' proportion of the lead consumed in the US leaped from 50 per cent to 85 per cent over this same period. The cathode ray tubes in our televisions and computer screens, which contain as much as two kilograms of lead each, have added to growth in lead usage. At the same time, all this lead is far from being safe for many outside the US from those working or living around where it is smelted and processed, to those exposed via a blossoming international trade in electronic waste.[16]

Americans' public health victory over lead reflected in those charts of declining blood lead levels has not come without costs. Even as Americans' and other Westerners' lives depend on lead as much as ever, its presence in our environment has been carefully constructed, materially as well as socially, for us not to give it another thought. Those contaminated toys from China may have surprised us, but more unawares, we had already become remote consumers of the produce of Chinese and other developing-world mines and smelters. Yet we have found it ever easier to lose sight of our dependence on, and complicity with, the continuing imposition of these hazards on victims elsewhere.

9.2 Multiple scales in a past time: the early twentieth-century 'global'

As in other arenas of history, once we seek to square what we know about lead today with what we can find out about earlier periods, we run into problems of evidence. While comparable historical statistics are now available for many nations' extractive and manufacturing industries, statistics available for earlier eras, even the early twentieth century, are considerably more fragmentary and difficult to compare. We may nevertheless begin with transnational vantage points becoming available for tracking the lead industries in various nations, namely, the statistics compiled early by the League of Nations. In its tracking of national efforts to recognise and regulate lead's dangers, a group like the ILO, founded in 1919 alongside the League of Nations, helps illuminate these cross-national patterns in the regulation of industrial hazards. Such international institutions and their 'global' perspectives deserve much further attention from industrial hazard historians. To understand better the 'global' of a group such as the ILO, to flesh out just what its perspective did and did not encompass, we need to scrutinise its constitution, meaning, and consequences not just within but beyond its Geneva headquarters, at a variety of scales and locales around the world.

The ILO coalesced only after decades-long transformations in lead's production and consumption in its chief supporting nations, mostly European and developed, accompanied by rising scrutiny of health experts and the state.[17] While a more recent understanding of lead has focussed on extra-workplace, environmental contamination, lead talk in the early ILO, reflecting the rise of labour unions, confined its purview to worker hazards at the site of production itself. The ILO's early 'conventions', recommended as 'universal' laws, often addressed particular industrial dangers faced especially by nations in the developed world. For instance, one of its first conventions, a 1921 ban on white lead pigments in interior paint, was far less concerned about the exposure of children or consumers, more with dangers to adult painters. Also characteristic was the ILO's reliance on a classic clinical presentation of lead poisoning, whose wide recognition helped make this one of the first three diseases recommended for workers' compensation.[18] A strict historicism, hewing only to the lead hazards that the ILO and its experts saw at the time, yields many interesting questions for which the organisation's archives in Geneva, voluminous and nearly untapped, promises abundant answers.[19] But we gain additional analytical leverage by recognising important limitations in how this international agency conceived of an early twentieth-century flow of lead into human societies, environments and bodies. What stands out, compared to today's concerns about lead, is just how little consideration the ILO devoted to contacts with lead outside the workplace. Everything else was considered 'public health', an enterprise at this time largely centring on germs.[20]

On this as with other questions, further clarity about the ILO's version of the 'global' comes when we shift our sights downward in scale, to what was happening in particular nations. The ILO had limited influence on industrial hygiene in the United States, the China of the early twentieth century in terms of being the world's largest lead producer. The US was also a nation where the unions were arguably weaker and less politically effective than those in many parts of Europe. American state-building in industrial health had come more slowly and hesitantly, and was far more decentralised. As significant, America's lead industries grew less because of multinationals moving in from abroad than through indigenous corporate growth, perhaps not so different from China's a century later.

Early American hygienists like Alice Hamilton drew on their own visions of the global, making workplace lead and its effects more publicly visible within the United States. Drawing on the literature and experience of other developed nations across the Atlantic, they challenged prevalent combinations of oversight and indifference in their own country: factory inspectors, American doctors, managers and owners who believed American factories and workers to be immune from European ills. By the 1920s the 'good progress' of American hygienists, based on a strengthening combination of factory and workers compensation laws, won them entry in the ILO's circle of international experts. Yet despite labour union pressure, the US refused to apply for ILO membership until the 1930s, and American states adopted ILO conventions only fitfully.[21] Some provided compensation for lead poisoning, others did not; none of them went so far as to ban white lead in paint. An abundance of lead came coupled with what, by ILO standards, was a relatively weak and porous regulatory system. Situating the ILO's 'global' over and against such national experiences emphasizes, among other things, how much the ILO's 'global' knowledge and prescriptions may have been gauged to the national politics and states, as well as lead exposures, that were more prevalent in Europe than in the United States.[22]

Developing-world industrial hazards remained marginal not just to the ILO's but also to the American hygienists' visions during this period. To chart this marginality, we also need to reach down to individual firms and the local level. While some lead-using firms and factories became subjected to American industrial hygienists' scrutiny, for instance, others did not. Most marginalised were those in the most rural and undeveloped, as well as least 'American' corners of the nation, at the border with its Spanish-speaking neighbour. Few lead smelters were less seen by the American state and its professional companions than one built in 1887 just outside of El Paso, Texas, by Phelps Dodge. Taking advantage of lead mines in northern Mexico, it lay not a hundred yards from the Mexico, just across the river from the Mexican town of Cuidad Juarez.[23] Hamilton, though she visited smelters in Arizona and Missouri, did not make it out to this most 'remote' of American smelters. It seems likely that workers in this smelter were like those found in a Missouri smelter Hamilton did visit: 'full of malaria, hookworm, and silica dust, from the chat heaps, to say nothing of lead'. By the early 1930s, the state in which it arose still had no workers compensation law for lead poisoning.[24] The continued exclusion of this El Paso smelter from regulatory regimes that were standard in other American states suggests an intra-national and regional race to the bottom in early twentieth-century America that anticipated the cross-national migrations of industrial hazards later on.

The example of the El Paso smelter also illuminates the industrial hygienists' working assumptions in this period about a certain level and style of urbanisation. In Chicago, where Hamilton began her work, a city health department tackled issues of sanitation, housing and sewage, not to mention smoke, justifying industrial hygiene's confinement to factory interiors. But along the Mexico-US border in this period, a place like 'Smeltertown' (see Illustration 9.1), right beside the El Paso smelter was subject to a steady barrage of lead and other fumes from its smokestack, with little or no intervention from any local health officer.

Illustration 9.1 The El Paso smelter and vicinity c. 1889
Source: El Paso Public Library.

We know from later studies that exposure in and around these homes could easily reach those in many parts of the plant.[25]

Finally, looking for lead across the border in Mexico itself illuminates assumptions about industrialising that further blinkered American hygienists to lead's actual course through theirs as well as this neighbouring nation. By the early 1930s, though American capital had funded an industrial scale of mining and smeltering there, many uses of lead there also remained closer to a craft scale of industry, with which the professional hygienists either in America or Europe rarely bothered. In villages of southerly Mexico, thousands of families earned income through pottery production that relied on leaded glazes for colour. Few uses of lead were more traditional than this, in which work exposures were barely separated from those inside the home. Most of their small open kilns lay within home lots. Interestingly, recent studies of these family potting enterprises find that both children and adults living in such a place get their biggest doses of lead from the food cooked in their leaded pots.[26]

We already have some sense of how blind eyes to these routes of exposure to lead may have shaped the global visions among America's industrial hygienists in this period, but less of how they may have governed the contrasting approaches to lead during the 1930s by Mexican versus American politicians. America's hygienists saw Mexico's industrial hazards as so thoroughly backward that lead researcher Robert Kehoe headed to Mexico in 1933 to establish primitive exposure levels. In tiny mountain villages around Ixtlahuaca, he sought a contrast with those he measured for non-occupational exposures in American cities. Among a population eating out of lead-glazed pots, he surmised he was finding a pre-industrial normal lead level.[27] Ironically, however, a worker and peasant-friendly Cardenas regime in Mexico had soon gone much further than did the American New Dealers in passing those lead-related 'conventions' or laws recommended by the ILO. In 1934, it made lead poisoning compensable nationwide, something that only America's most industrial states had done. Some four years later the Mexican national government passed a convention that never made it past any legislature in the US, outlawing white lead paint in interiors.[28]

Around this same time, change was afoot internationally in the ILO itself and soon thereafter through establishment of another Geneva-based organisation interested in industrial health. The US finally joined the ILO in 1934, helping shift its membership based beyond Europe. This international agency also began turning more attention to the plight of industrial hygiene in the developing world, through surveys including one of training programmes in industrial medicine and hygiene. Compensation for a disease like lead poisoning, after all, arguably meant little if no knowledgeable eyes were around to recognise or confirm it; not surprisingly, there were far more such programmes in the US than anywhere in Latin America.[29] In taking up this concern, the ILO was joined by other new post-World War II international institutions, notably the WHO. Simultaneously, new turns in the knowledge and regulation of toxins such as lead, based in places like the United States, bred new cross-national contrasts in approach, as well as actual exposures, to industrial hazards.

9.3 The uneven development of post-war precautionism

Again, most of historical scholarship about the more aggressive approaches to lead regulation emerging after World War II has confined itself to events within scholars' own nations. The recent burst of American scholarship shows how a more precautionary approach to lead came to be forged, especially on exposures beyond the workplace, via markets as well as what became known as 'environment'. The chief usages of concern were less those inside factories as those by consumers, from peeling lead paint to the burning of tetraethyl lead in gasoline.[30] Among the scientific changes ushering it in, over the 1950s into the 1970s, ever more subtle and long-term effects of lead poisoning were established especially among children: behavioural disturbances, slowed nerve conduction and lowered IQs. Scrutiny of lead levels in places where no one lived, such as the Greenland ice shelf, also showed pre-industrial lead exposures turned out to be several times lower than what Kehoe had found in his Mexicans.[31] Changes in lead science and regulation in the post-World War II US owed much to the rise of a federal and academic research establishment devoted to chronic degenerative ailments such as cancer and heart disease, now widely recognised as the primary killers of Americans. Public funding sources for research into lead and other industrial and environmental hazards replaced an earlier reliance on corporate funding, exemplified by Kehoe. After innovations in epidemiology and toxicology helped establish why a new level of precautionism was necessary, new federal laws and regulation took on a whole new range of consumer and environmental hazards, from food additives to cigarettes to industrial pollution.[32] For all we are now learning about the history of this precautionism within particular nations like the US, the cross-national dynamics that also drove or facilitated its advent have received less attention. As with the preceding period, study of the 'global' visions of the time, like that of the ILO, may reveal much; so too, that of carefully selected local sites.

Among the tides conveying many nations toward precautionism over the post-war period was the widespread adoption of the Maximum Allowable Concentration (MAC) levels by their industrial hygienists. This strategy, first used in a tentative way by French, German and English investigators, came to full fruition in the industrial hygiene circles of the United States and also Russia over the 1930s and 40s.[33] It reflected a growing confidence in engineers' ability to monitor and control the atmospheric concentration of toxic dust such as lead, rather than waiting until workers actually became sick. Broaching the possibility that clinical poisoning could indeed be avoided, they cracked open the door to still more preventive approaches. We have much to learn about how countries set about adopting their own lists of MACs, as well as how the ILO and WHO sought to harmonise these approaches into a single international standard.[34] One driver of change appears to have been the Cold War, which nourished a competitive debate over how lead hazards should be studied and abated. From the start of MACs in the 1940s, Americans, following epidemiological and clinical studies by the US Public Health Service, set their tolerance levels for lead dust where clinical symptoms started to occur. The Russians led their Eastern European allies in urging stricter standards, more like those in the American regulation of food additives, based on toxicological experiments with animals. Despite American dismissals of post-war Russian research claims as propaganda (and their comparable neglect in the actual factories of Russia and Eastern Europe), these turned out to be dead-on predictive of where US lead research would go over the 1960s and 1970s: into its effects on higher nervous activity.[35]

While these international debates over lead science and regulation were no doubt important, looking at border-straddling industrial locales, like the El Paso smelter can illuminate a great deal about differing roots and implications for the new precautionism in a developed, versus a less developed nation. While the science moved beyond the worker epidemiology underpinning the early MACs, by no means was it exclusively focused on consumer exposures like that to tobacco. Long-neglected or under-regulated factories such as El Paso's lead smelter served as pivotal sites for forging a new environmental epidemiology of residential populations. Phillip Landrigan, a paediatrician working for the Centers for Disease Control, travelled to Smeltertown and the nearby vicinity in 1971, to undertake detailed environmental and clinical study of lead exposures in the community surrounding the factory. Among neighbourhood children, his team found strong, statistically significant correlations between lead exposure and blood levels of more than 40 micrograms per decilitre, and similarly robust connections between this level of absorption and brain or nerve impairment. This study was cited as providing some of the most conclusive evidence for the falling lead standards of the late 1970s for workplace as well as outdoor air.[36]

Looking at the local history not just of this factory but of its surrounding environment illustrates how much more was at work in this new precautionism than professional and scientific change. What enabled the new laws, agencies and funding that fuelled precautionism toward lead and other hazards was not just new science but a groundswell of public concern about industrial pollution. While labour unions and work stoppages played a role in this explosion of concern in the US, it flourished especially among those living in suburbs.[37] Trends in the larger county where the El Paso smelter was located point to a widespread presupposition about urbanising that made the new precautionism possible: that it had become possible to live as well as work at a comfortable, less polluted remove from such industrial plants. Over the post-war period, El Paso's well-to-do were moving away from the factory area (Figure 9.3). Smeltertown itself came to be inhabited especially by

194

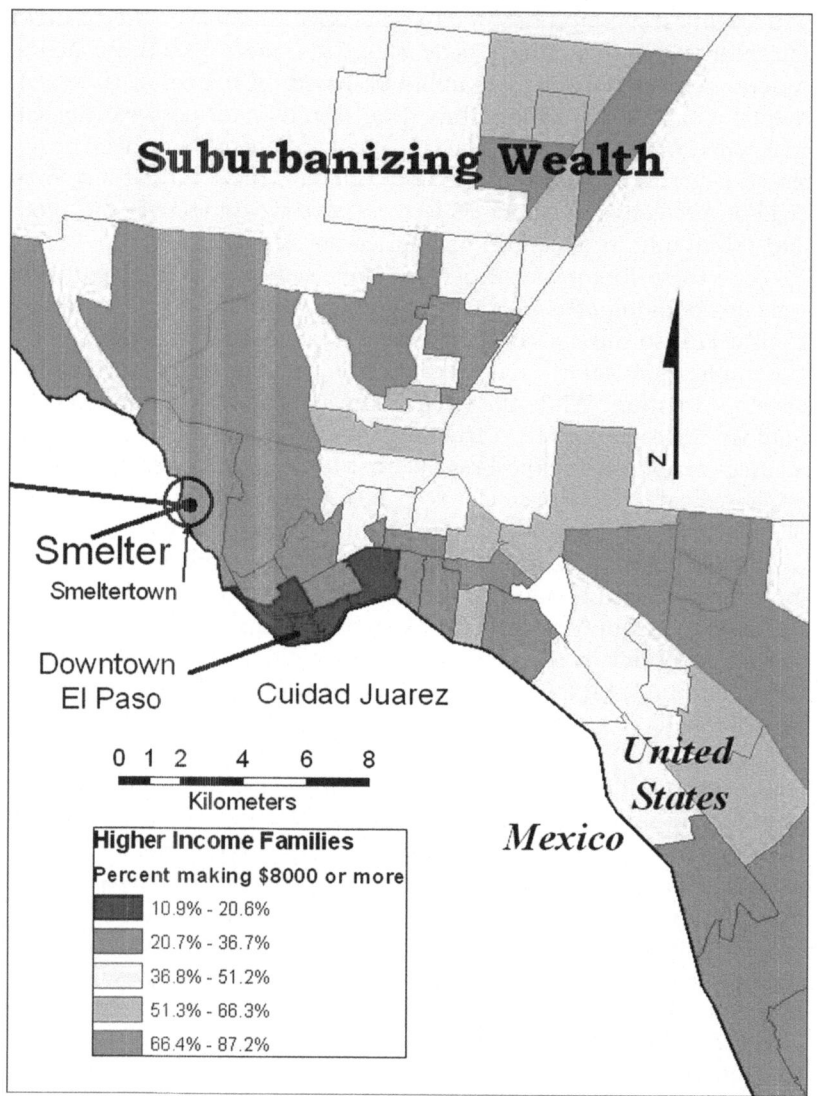

Figure 9.3 Higher income families by census tract, 1970 in the El Paso area
Data Source: http://www.nhgis.org/ Map by Christopher Sellers.

lower income families of Hispanic origin, whose breadwinners dominated the smelter's workforce. Those moving to the suburbs were more
likely to be wealthy and white, and less likely to do smelter work. The
new public support for precautionism reflected an evolving urban
geography in the post-World War II United States, marked by contrasts
in ethnicity or race, occupation and wealth, which held out the possibility of total escape from the factory. Of course, in a place like El Paso,
industrial pollution also denied it, by also finding its way into suburbs
that were whiter and wealthier. The Landrigan study was triggered by
just such exposures and their political repercussions.[38] Today the model
for industrial pollution depicted on the Environmental Protection
Agency (EPA) website captures this 'post-industrial' detachment from
the factory in an idealised form (Figure 9.4). Nobody works or lives
around these polluting factories; everyone lives and works across the
river. Here the provision of running water, flush toilets, and water treatment, those sanitary amenities pushed by earlier generations of health
authorities, remain invisible, taken for granted.

If these local urbanising trends undergirded this new precautionism
on the American side of this border, trends in industrialisation took
the Mexican side in a quite different direction. On a more global scale,
factories had begun migrating out of developed nations like the United
States to developing countries like Mexico. The Mattel Corporation
itself offers a nice example. Begun in the Los Angeles area just after
World War II, by the early 1960s its southern California factories had
undergone crippling strikes and it was in search of the spatial fix of
new factory settings. Among the first American companies to set up a
factory in China, it also turned to Mexico, whose government in 1965
set up tariff-free zones for American companies just south of the US
border. While Mattel and other toymakers chose Tijuana, helping bring
it the biggest concentration of these 'maquiladora' operations, El Paso's
neighbour Ciudad Juarez was next in line by 1986, with nearly 200.

With this industrialising came an urbanisation of quite different
intensity and character from that on the American side. Whereas
between 1940 and 1980, El Paso grew fivefold, Ciudad Juarez, on the
other side of the Rio Grande, expanded twenty-fivefold. As the US EPA
was created and began to implement new federal environmental policies, Cuidad Juarez's population growth was rapidly outstripping that
of El Paso, to attain nearly half a million by 1980. Along with Tijuana
just south of San Diego, where Mattel's biggest Mexican operation lay,
these cities emerged as Mexico's biggest industrial metropolises outside
of Mexico City.[39] The pattern of city growth contrasted even more

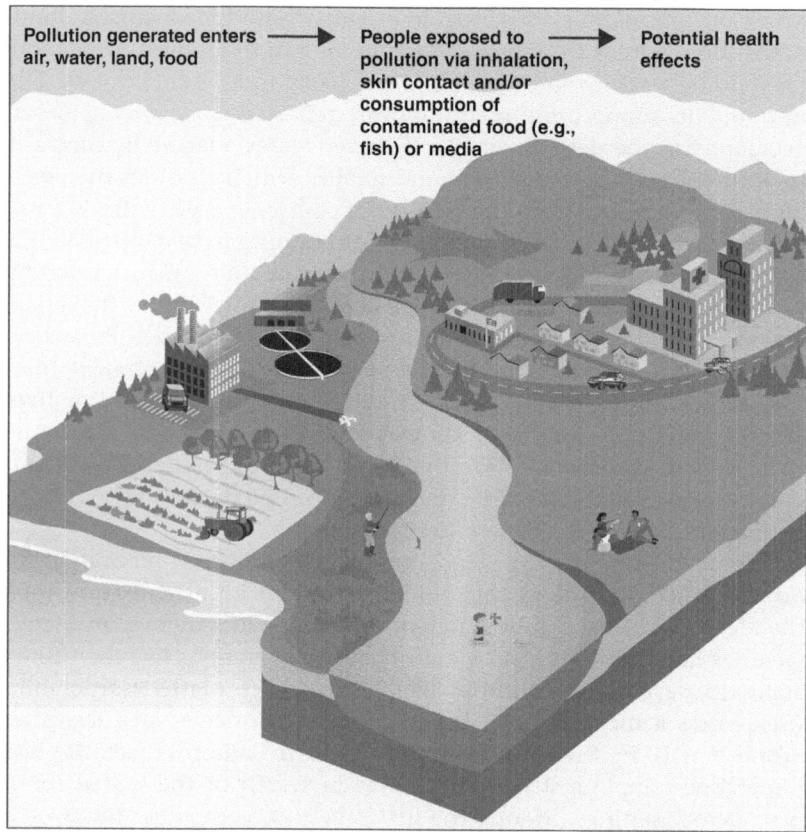

Figure 9.4   Idealised depiction on the EPA website, 2004. The factory and other sources of pollution lie across the river from all depicted residences
Source: http://www.epa.gov/indicate/roe/pdf/EPA_Draft_ROE.pdf.

starkly with that on the American side of border, where precautionist science and principles had begun to flourish. Instead of moving further out from Smeltertown, as did El Pasans, those newcomers living in and around Ciudad Juarez were crowding closer in. Just across from the factory there emerged one of the colonias through which Cuidad Juarez and other Mexican towns grew. An aerial photo from 1982 shows a stark contrast between the suburban neighbourhoods and other settlements to the north, with small houses and bare yards, without lawns.[40]

These contexts suggest reasons why the embrace of precautionism may have come slower in Mexico than it did in the US, even as the exposures highlighted by the new environmental science were indeed worse. By the 1970s, Mexican health experts quickly undertook their own counterpart studies to the environmental lead investigations of Landrigan, duplicating his epidemiological approach. Their investigation was smaller, with no accompanying neurological studies. Yet they found more children at risk, including 52 per cent of those within a mile of the smelter.[41] That the lead exposures were correspondingly more massive was, in part, a reflection of settlement patterns that brought many more children into closer range of the plant. Mexican public health officials in a town like Juarez nevertheless faced more difficult choices than their American counterparts in prioritising, for lead poisoning was not the only or even the most dangerous health threat to those living in places like the colonias. As early as 1977, the *New York Times* was reporting hordes of squatters on Juarez's outskirts, including many families with children. In addition to dirty air, many of these houses were also without the sanitary safeguards like running water and flush toilets that Americans largely took for granted. Sewage ran through open channels paralleling its streets. Monitoring of the water supply for bacteria only began in the 1990s on the Mexican side, and hepatitis and tuberculosis remained consistent problems.[42]

If the new precautionist experts and regulations to the north furnished their Mexican counterparts with ready-made methodologies and remedies, they faced these as well as other dilemmas that did not trouble American precautionists. For instance, following US as well as other international examples, Mexico's federal government inaugurated a shift away from leaded gasoline in 1986. By 1998 lead had been eliminated from all gasoline sold in the Mexico City area and mean aerial lead levels were plummeting. Yet the ways in which Mexico's national oil company Pemex reformulated gasoline to take the lead out combined with this megacity's predominantly older car fleet to exacerbate other kinds of air pollution, notably from ozone. In the realm of lead, as well, Mexico's would-be precautionists still confronted an array of endemic exposures their American counterparts did not. Forty per cent of families living in this metropolitan area still used lead-glazed ceramics. Outside the city, not only was lead pottery usage still more pervasive, a rising tourist trade in this kind of ceramics has enabled 1.5 million Mexican families to continue to rely on family kilns for income.[43]

By 2007, when the story of Mattel's toxic toys erupted, it is a safe, if ironic, bet that where the precautionist way of seeing lead had the biggest impact in the whole of Mexico was probably in Mattel's plant in Tijuana. Since Mattel shut down its last plant in the United States in 2002, its testing facilities for what it sells there have been confined to this south-of-the-border factory. That is where Mattel quietly began shipping the China-made goods still in its American warehouses, once the story of contamination broke. In Mexico City, meanwhile, consumer testing of toys by the national government only began in 1994, nearly two decades after Mattel arrived in Tijuana.[44] And when it comes to toys, the Mexican consumer agency faces another level of dilemma from its American or European counterpart: a huge market of native-made knock-offs or contraband. Nearly half the toys sold are illegal, hence, not reachable even by stepped up testing. Those toys that Mattel labs reject for their lead content may well wind up being sold by unlicensed street-side vendors.[45]

We Americans and European have come to precautionism only through decades, even a century, of institutional development in our governments and professions, also more quietly, in our formal private economy. In America, a new self-described environmental or ecological approach to industrial hazards coalesced hand-in-hand with transformations of American cities, homes and jobs, which enabled people to imagine factories, as well as public health interventions like DDT, as doing more harm than good. No doubt there were, and still are, corners of America and Europe untouched by precautionism, where the more subtle risks of lead exposure are ignored or sloughed off. But more dramatic differences in how lead is seen and interpreted – as well as actually distributed – are to be found across a nation such as Mexico. In Mexico, as in many parts of the developing world, experts and advocates not only confront a greater magnitude and variety of lead hazards, these come piled on top of others, including sanitary ones their US counterparts have long since faced and conquered. In Ciudad Juarez, industrial and childhood lead poisoning have multiplied, and become recognised, hand-in-hand with water-borne infections. There as in Mexico City, lead exposure has proven difficult to tame by those measures that worked in the United States. Where by best-guess projection, lead's human contacts remain so much more pervasive and persistent, the popular as well as medical cultures of lead cannot but be different from those in the states.

9.4 Conclusions

Study of the genesis of these differences offers many engaging possibilities for medical, health and other historians. We may, for instance, shed light on resistances to public health interventions which health activists have confronted in a place such as Mexico.

A more robust attunement to cultural as well as environmental differences, and the longer historical contrasts that lie in back of them, may help us better understand those contrasting popular categories through which lead and its exposures are understood. With such an understanding may come further appreciation of just what may stand in the way of inculcating a more modern lead ontology, those categories, values and contexts that stir active resistance to the health educators' messages. What seems clear is that the opposition and scepticism faced by Mexican advocates runs deeper than the resistance of multinational corporations, or the mistaken beliefs pinpointed by health educators. Our own 'modern' prescriptions for lead reach into what many Mexicans see as new economic opportunities, but also into traditional spheres of knowledge and practice. Either way, they implicate ways of life and livelihood that are not, or are no longer shared by most Americans and Western Europeans.

Along this and other fronts, the cross-national study of industrial hazards has much to offer. Especially if we are willing to range across the available intellectual registers, to strive after an environmental as well as a socio-cultural symmetry in our analysis, we can understand simultaneous trajectories of nations as different as the US and Mexico in comparable terms. We thereby also begin to fathom just how asymmetrical their trajectories have been. Differing histories of the engineering as well as well control of lead's flows and exposures undergird contrasting ontologies of lead and its effects in these two nations. Central to the achievement of footloose firms like Mattel has been their ability to extract profit from such cross-national differences in the ontology of toxins like lead, while keeping their own hands relatively clean. Until very recently, perhaps.

Through the last several decades, the flight of factories and industrial production from the developed to the developing nations has reshaped our world, but also subtly warped how medical and other historians have seen it. Simply by tackling this important subject, we join many other scholars in helping to puncture conceits such as that ours is a 'post-industrial society'. What are such notions, if not ways for developed-world scholars to avert our eyes from our continuing dependencies on production elsewhere? Within the scholarship that challenges such ideas, health and medical as well as environmental historians have an especially critical role to play. We come intellectually armed to bring out what many other scholars, perhaps by dint of other modes of insulation, have tended to neglect: that this historical shift in the geography of industry has had human bodily as well as ecological consequences. Among the rewards in taking on historical contrasts as marked as those between the United States and Mexico, is the deeper and broader understanding of what has enabled health advances in a nation such as United States itself. Through such comparisons, the underlying conditions for public health victories like that of America's precautionists over lead – from the human capital of expertise, the eradication of pre-existing 'traditional' materials and practices, the peculiar style of post-World War II urbanising – stand out precisely through their relative absence in a places such as Mexico, where precautionists have had a considerably harder time.

Most fully to plumb the gravity and implications of such a topic, however, we cannot afford an agnosticism towards the various sciences of lead that have emerged across time and place; on what the dangers of lead are, we must take ontological sides. After all, what appeals in this, as so many other topics that today's historians choose to take up, is its moral dimensions, what historical investigation may say to our own place and time. Such a transnational history reinforces the message that such toxins' currents in a place such as Mexico or China may continue to track back to us, those in the developed world, who now ordinarily imagine ourselves living 'lead-free'. In an important sense, we reconnect with their distant smelters and their sickened workers and their children each and every day, whenever we start up our cars or turn on our computers. We do not so much through our perception of lead, nor through any bodily hazard it poses to us. We do so, instead, by reliance on the lead which they have passed our way. This common material thread also binds us to the many hazards and victims it has accumulated during its passage. Building this commonality into our histories invites a search for ways of taking responsibility for the costs it imposes, however distant and remote, even if they are not our nation's own.

Notes

1. 'Millions of mattel toys recalled in Europe', *Germany/UK* (14 August 2007) at http://tvscripts.edt.reuters.com/2007-08-14/1aa2f5a9.html, accessed 23 August 2007; 'Parents urged to check homes as 2 million unsafe toys are recalled', *Timesonline* (15 August 2007) at http://www.timesonline.co.uk/tol/news/uk/article2260555.ece, accessed 23 August 2007; 'Mattel Inc. toys in Asia recalled on concerns over lead in paint', *International Herald Tribune/Business* (3 August 2007) at http://www.iht.com/articles/ap/2007/08/03/business/AS-FIN-Asia-Toy-Recall.php, accessed 5 May 2008. A briefly noted exception was China Labor Watch, 'Investigation on toy suppliers in China; workers are still suffering', (21 August 2007) at http://www.chinalaborwatch.org/20070821eighttoy.htm, accessed 30 May 2008.

2. K. Marx (1967). *Capital*, vol. 1(New York: International Publishers), pp. 71–83.
3. H. E. (Henry Ernest) Sigerist (1891–1957) (1936). 'Historical background of industrial and occupational diseases', *Bulletin of the New York Academy of Medicine* 12, 2nd, pp. 597–609; L. Teleky (1948). *History of Factory and Mine Hygiene* (New York, Columbia Univ. Press); G. Rosen (1943). *The History of Miners' Diseases, a Medical and Social Interpretation* (New York: Schumans).
4. Earliest work includes Paul Weindling and Society for the Social History of Medicine, (1985). *The Social History of Occupational Health* (London; Dover, N.H.: Croom Helm); D. Rosner and G. Markowitz (1987). *Dying for Work: Workers' Safety and Health in Twentieth-Century America* (Bloomington: Indiana University Press); the latest includes: A. McIvor and R. Johnston (2005). 'Medical knowledge and the worker: occupational lung diseases in the United Kingdom, c. 1920–1975', *Labor: Studies in Working Class History of the Americas*, 2, no. 4, pp. 63–86; M. W. Bufton and J. Melling (2005). '"A mere matter of rock": organized labour, scientific evidence and British government schemes for compensation of Silicosis and Pneumoniosis among coal miners, 1926–1940', *Medical History*, 49, no. 2, pp. 155–78; A. Vergara (2005). 'The recognition of Silicosis: labor unions and physicians in the Chilean copper industry, 1930s–1960s', *Bulletin of the History of Medicine*, 79, no. 4, pp. 723–48.
5. The recent literature is well-illustrated by a special issue of the journal *Environmental History*: J. A. Roberts and N. Langston (2008). 'Toxic bodies/toxic environments: an interdisciplinary forum', *Environmental History*, 13, no. 4 at http://www.historycooperative.org.libproxy.cc.stonybrook.edu/journals/eh/13.4/roberts.html, accessed 9 May 2009; also S. Mosley (2006). 'Common ground: integrating social and environmental history', *Journal of Social History*, 39, no. 3, pp. 915–33.
6. On various symmetry principles invoked in science studies, see D. Breslau (2000). 'Sociology after humanism: a lesson from contemporary science studies', *Sociological Theory* 18, pp. 289–307; A. Preda (1999). 'The turn to things: arguments for a sociological theory of things', *Sociological Quarterly*, 40, pp. 347–66.
7. The phrase 'matters' from J. Butler (1993). *Bodies That Matter: On the Discursive Limits of 'Sex'* (New York and London: Routledge); see also the literature on historical ontology and related methods: I. Hacking (2002). *Historical Ontology* (Cambridge: Harvard University Press); L. Daston, (2000). *Biographies of Scientific Objects*, 1st edn (University Of Chicago Press); D. Ihde and E. Selinger (eds) (2003). *Chasing Technoscience: Matrix for Materiality* (Bloomington: Indiana University Press); also C. Sellers (2004). 'The artificial nature of fluoridated water: between nations, knowledges, and material flows', in, G. Mitman, M. Murphy and C. Sellers (eds). *Osiris – Landscapes of Exposure; Knowledge and Illness in Modern Environments*, 19, pp. 182–200.
8. A. D. Chandler Jr (1994). *Scale and Scope: The Dynamics of Industrial Capitalism* (Belknap Press of Harvard University Press); M. D. Bordo, A. M. Taylor and J. G. Williamson (2005). *Globalization in Historical Perspective* (University Of Chicago Press); I. M. Wallerstein (2004). *World-Systems Analysis: An Introduction* (Durham: Duke University Press); B. J. Silver (2003). *Forces of Labor: Workers' Movements and Globalization Since 1870* (Cambridge University

Press); D. Harvey (1991). *The Condition of Postmodernity: An Enquiry into the Origins of Cultural Change*, Reprint (Wiley-Blackwell); D. Harvey (1997). *Justice, Nature and the Geography of Difference* (Wiley-Blackwell).

9. D. Harvey (1981). 'The spatial fix: Hegel, Von Thunen, and Marx', *Antipode* 13, no. 3, pp. 1–12; W. Greider (1997). *One World, Ready or Not: The Manic Logic of Global Capitalism* (New York: Simon & Schuster); J. R. Cowie (1999). *Capital Moves: RCA's Seventy-Year Quest for Cheap Labor* (Ithaca, NY: Cornell University Press); Silver, *Forces of Labor*.

10. J. R. McNeill, (2001). *Something New Under the Sun: An Environmental History of the Twentieth-Century World* (W.W. Norton & Co.); I. Tyrrell (1999). *True Gardens of the Gods: Californian-Australian Environmental Reform, 1860–1930*, 1st edn (University of California Press); J. Soluri (2006). *Banana Cultures: Agriculture, Consumption, and Environmental Change in Honduras and the United States* (University of Texas Press); J. L. Anderson (2007). *Nature's Currency: The Atlantic Mahogany Trade, 1720–1830*.

11. R. White (2004). 'From wilderness to hybrid landscapes: the cultural turn in environmental history', *Historian*, 66, no. 3, pp. 557–64; Roberts and Langston, 'toxic bodies/toxic environments'; G. Mitman, M. Murphy and C. Sellers (eds). *Landscapes of Exposure: Knowledge and Illness in Modern Environments*, 19, *Osiris* 2 (Chicago: University of Chicago Press, 2004).

12. J. O. Nriagu (1990). 'Global metal pollution (cover story)', *Environment*, 32, 7, pp. 6–11, doi: Article; J. O. Nriagu (1998). 'Tales told in lead', *Science* 281, no. 5383, new series: pp. 1622–23.

13. 'The burden of occupational illness', Press Release WHO/31 (8 June 1999) at www.who.int/inf-pr-1999/en/pr99-31.html, accessed 3 June 2008.

14. For instance, A. Chan and R. J. S Ross (2003). 'Racing to the bottom; international trade without a social clause', *Third World Quarterly* 24, 6, pp. 1011–28; J. Simpkins (2004). The global workplace: challenging the race to the bottom', *Development in Practice* 14, no. 1/2, pp. 110–18; Greider, *One World Ready or Not*.

15. For one of these charts, see the following http://www.epa.gov/bns/lead/Fig_01.gif, accessed 5 February 2010.

16. M. B. Biviano, L. A. Wagner, D. E. Sullivan (1999). *Total Materials Consumption; An Estimation Methodology and Example Using Lead; A Materials Flow Analysis*, US Geological Survey Circular # 1183 (Washington, DC: US Govt. Print. Off).

17. W. Martin (1926). 'The International Labour Organisation', *Proceedings of the Academy of Political Science in the City of New York* 12, 1, pp. 399–410; M. Ruotsila (2002). '"The great charter for the liberty of the workingman': labour, liberals and the creation of the ILO', *Labour History Review*, 67, 1, pp. 29–47; J. VanDaele (2005). 'Engineering social peace: networks, ideas, and the founding on the International Labour Organization', *International Review of Social History*, 50, 3, pp. 435–66.

18. J. Heitmann (2004). 'The ILO and the regulation of white lead in Britain during the interwar years: an examination of international and national campaigns in occupational health', *Labour History Review*, 69, 3, pp. 267–84; M. Robert and L. Parmeggiani (1969). *Fifty Years of International Collaboration in Occupational Safety and Health*, vol. 19, CIS information sheet (Geneva, International Occupational Safety and Health Information Centre); International Labour

Office (1930). *Occupation and Health; Encyclopedia of Hygiene, Pathology, and Social Welfare* (Geneva).

19. For an instance of what can be done, see T. Cayet, P.-A. Rosental and M. Thébaud-Sorger, (2009). 'How onternational organisations compete occupational safety and health at the ILO, a diplomacy of expertise', *Journal of Modern European History*, 7, no. 2, pp. 174–96; for an overview, see the archives website, at http://www.ilo.org/public/english/century/information_resources/ilo_archives.htm, accessed 3 February 2010.

20. International Labour Office, *Occupation and Health; Encyclopedia of Hygiene, Pathology, and Social Welfare*, 'Homework', pp. 964–70; 'Industrial waste water (Treatment of)', pp. 41–9.

21. C. C. Sellers (1997). *Hazards of the Job: From Industrial Disease to Environmental Health Science* (Chapel Hill: University of North Carolina Press); V. Jobst III, (1938). 'The United States and international labor conventions', *The American Journal of International Law*, 32, pp. 135–8; C. J. Ratzlaff (1932). 'The International Labor Organization of the League of Nations: its significance to the United States', *The American Economic Review*, 22, 3, pp. 447–61.

22. Workmen's compensation for occupational diseases. Partial revision of the convention [...]. report V. International Labour Conference, Eighteenth Session, Geneva, 1934. (Geneva: International Labour Office); Factory Inspection (1923). *Historical Development and Present Organisation in Certain Countries* (Geneva: International Labor Office).

23. J. W. Drexler, 'finalreport.pdf (application/pdf Object)', *A Study on the Source of Anomalous Lead and Arsenic Concentrations in Soils from the El Paso Community – El Paso, Texas*, at http://www.epa.gov/region6/6sf/pdffiles/finalreport.pdf, accessed 6 June 2008; I. F. Marcosson (1949). *Metal Magic; the Story of the American Smelting & Refining Company* (New York: Farrar, Straus).

24. A. Hamilton (1985). *Exploring the Dangerous Trades: The Autobiography of Alice Hamilton, M.D.* (Boston, Mass.: Northeastern University Press), p. 146 (quote); A. Hamilton (1914) *Lead Poisoning in the Smelting and Refining of Lead*, vol. 141, Bulletin of the United States Bureau of Labor Statistics (Washington: G.P.O.); *Workmen's Compensation for Occupational Diseases. Partial Revision of the Convention [...]. Report V. International Labour Conference, Eighteenth Session, Geneva, 1934)*, p. 182.

25. Hamilton, *Exploring the Dangerous Trades*, for ex., pp. 97–100; P. J. Landrigan et al. (1975). 'Epidemic lead absorption near an ore smelter. The role of particulate lead', *New England Journal of Medicine*, 292, 3, pp. 123–9; B. R. Ordóñez, L. Ruiz Romero and I. R. Mora (1976). 'Investigacion epidemiológica sobre niveles de plomo en la población infantil y en el medio ambiente domiciliario de ciudad Juarez, Chihuahua, en relación con una fundición de El Paso, Texas,' *Boletín de la Oficina Sanitaria Panamericana*, 80, 4, pp. 303–17; also the impetus for the first study came from a collaboration between the Centers for Disease Control and local health department officials.

26. G. M. Foster (1948). 'The folk economy of rural Mexico with special reference to marketing', *Journal of Marketing*, 13, 2, pp. 153–62; E. B. Sayles (1955). 'Three Mexican crafts', *American Anthropologist*, 57, 5, new series, pp. 953–73; R Hibbert et al., 'High lead exposures resulting from pottery

production in a village in Michoacán State, Mexico', *Journal of Exposure Analysis and Environmental Epidemiology*, 9, 4, pp. 343–51, doi: 10489159.

27. J. Cholak, R. A. Kehoe (1933). 'Lead absorption and excretion in primitive life', *Journal of Industrial Hygiene*, 15, pp. 257–72; in fairness to Kehoe's team, they did note and try to measure the influence of lead-glazed pottery.

28. See entries for 'Mexico' at 'ILOLEX: country information', http://www.ilo.org/ilolex/english/newcountryframeE.htm, accessed 11 June, 2007.

29. L. Carozzi (1939). 'Training in industrial medicine', *International Labour Review*, 40, 6, pp. 733–67; L. Carozzi and others, 'Folder: industrial hygiene Egypt – collaboration of ILO official with PH authorities', correspondence, Industrial Hygiene, International Labour Organization Archives.

30. C. Warren (2000). *Brush with Death: A Social History of Lead Poisoning* (Baltimore, MD: Johns Hopkins University Press); D. Rosner and G. Markowitz (2002). *Deceit and Denial: The Deadly Politics of Industrial Pollution* (Berkeley, CA: University of California Press); P. C English (2001). *Old Paint: A Medical History of Childhood Lead-Paint Poisoning in the United States to 1980* (New Brunswick, NJ: Rutgers University Press).

31. C. C. Patterson (1965). 'Contaminated and natural lead environments of man', *Archives of Environmental Health*, 11, pp. 344–60.

32. An early confrontation came at the US Public Health Service, *Symposium on Environmental Lead Contamination*, vol. 1440, Public Health Service Publication (Washington, DC: U.S. Dept. of Health, Education, and Welfare, Public Health Service, 1966).

33. On early uses by French, German and English, see 'Air-testing in workshops', pp. 81–9 in International Labour Office, Occupation and Health; Encyclopedia of Hygiene, Pathology, and Social Welfare; on American conversion of MAC's into systematic tool, see C. Sellers, (2007). '"A prejudice that may cloud the mentality": making objectivity in early twentieth century occupational health', in J. W. Ward and C. Warren (eds) *Silent Victories: The History and Practice of Public Health in Twentieth-Century America*. (Oxford; New York: Oxford University Press).

34. International Union of Pure and Applied Chemistry, Proceedings of the International Symposium on Maximum Allowable Concentrations of Toxic Substances in Industry, Held in Prague, Czechoslovakia, April 1959 (London, 1961); C. Sellers, 'The cold war over the worker's body', paper prepared for Carcinogens, Mutagens, etc. Conference in Strasbourg, France, March 2010.

35. Ibid.

36. P. J. Landrigan et al., 'Epidemic lead absorption near an ore smelter. The role of particulate lead'; P. J. Landrigan et al. (1976). 'Increased lead absorption with anemia and slowed nerve conduction in children near a lead smelter', *The Journal of Pediatrics*, 89, 6, pp. 904–10; National Institute for Occupational Safety and Health, *Occupational Exposure to Inorganic Lead: Revised Criteria, 1978.*, vol. 78, DHEW (NIOSH) publication ([Cincinnati]: US Dept. of Health, Education, and Welfare, Public Health Service, Center for Disease Control, National Institute for Occupational Safety and Health, 1978).

37. H. Erskine (1972). 'The polls: pollution and its costs', *The Public Opinion Quarterly*, 36, 1, pp. 120–35; this is a central argument of C. C. Sellers, *Unsettling Ground: Suburban Nature and Environmentalism in Twentieth-Century America* (forthcoming from University of North Carolina Press).

38. 'Epidemiologic notes and reports on human lead absorption – Texas'.
39. L. Story, 'After stumbling, mattel cracks down in China', *The New York Times* (29 August 2007), at http://www.nytimes.com/2007/08/29/business/worldbusiness/29mattel.html?pagewanted=2&_r=1&sq=mattel%20and%20tijuana%20and%20lead&st=nyt&scp=1&adxnnlx=1212852280-giZjX6B1R8d%20OxWxz/n1sA, accessed 7 June 2008; T. Golden, 'New ball game for Mexican toys', *New York Times*, 28 August 1992, at http://proquest.umi.com/pqdweb?did=116253503&Fmt=7&clientId=48296&RQT=309&VName=HNP, accessed 6 August 2008; P. L. Martin (1992). 'Foreign direct investment and migration: the case of Mexican Maquiladoras', *International Migration* 30, 3, pp. 399–422; P. Cooney (2001). 'The Mexican crisis and the Maquiladora boom: a paradox of development or the logic of neoliberalism?', *Latin American Perspectives*, 28, 3, pp. 55–83; T. C. Brown (1997). 'The fourth member of NAFTA: The US-Mexico border', *Annals of the American Academy of Political and Social Science*, 550, pp. 105–21.
40. Aerial view of Smeltertown Area, 1982, show suburbs to the north and west of Smeltertown, in the US, with a very different pattern of urbanizing than those directly south of the border, in Cuidad Juarez and neighboring settlements. Source: www.lib.utexas.edu/maps/us_mexico_border/, accessed 3 February 2010.
41. J. Morales (2003). 'History of the national federation of occupational (FENASTAC), Mexico', in A. Grieco (ed.). *Origins of Occupational Health Associations in the World*, 1st edn (Amsterdam: Elsevier), esp. pp. 124–25; Ordóñez, Romero, and Mora, "[Epidemiological study of lead levels.]."
42. 'El Paso smelter still poses lead-poisoning peril to children in Juarez', *New York Times*, 28 November 1977, at http://proquest.umi.com/pqdweb?did=89692611&Fmt=7&clientId=48296&RQT=309&VName=HNP, accessed 8 June 2008; R. Suro, 'Border boom's dirty residue imperils US-Mexico trade', *New York Times*, 31 March 1991, at http://proquest.umi.com/pqdweb?did=115760549&Fmt=7&clientId=48296&RQT=309&VName=HNP, accessed 6 June 2008; Brown, 'The fourth member of NAFTA: The US-Mexico border'.
43. L. Schnaas et al. (2004). 'Blood lead secular trend in a cohort of children in Mexico City (1987–2002)', *Environmental Health Perspectives* 112, 10, pp. 1110–15; 'Working group III; ceramic glazes', C. Howson, M. Hernandez-Avila and D. Rall (eds) (1996). *Lead in the Americas: A Call for Action* (Washington, DC, USA: Committee to Reduce Lead Exposure in the Americas, Board on International Health, Institute of Medicine, in collaboration with the National Institute of Public Health), p. 129.
44. Story, 'After stumbling, Mattel cracks down in China'; '"Careful with the toys"', *ISI Emerging Markets Africawire*, 30 August 2007.
45. Ibid.

10

The Gardener in the Machine: Biotechnological Adaptation for Life Indoors

Christian Warren

Environmental historians, especially those working at the intersection of environment and health, talk about 'human interactions with the built environment'; this is, strictly speaking, redundant, since, by and large, what has defined the species *homo sapiens* since its appearance 30,000 years ago is our steadfast manipulation of the environment, cleverly exploiting resources and altering surroundings to counteract deficiencies in brawn, speed or other innate physical abilities.[1] Every technology we have adopted, from fire to stone tools, to agriculture and medicine, has enhanced our odds against the lions, tigers and bears. But we grew more dependent upon those technologies and came to rely less and less upon our bodies, our instincts and our knowledge of the natural world. The pace of this transformation has accelerated in the last two centuries, so that by the mid-twentieth century, careful observers worried that technological innovation was outstripping biological evolution in determining the nature of human existence, risking physical, social and psychic alienation from the natural order.[2]

Man is certainly not nature's only home builder, but it would seem that we are one of nature's more enthusiastic homebodies. In the modern imagination, our club-wielding caveman ancestors always dragged their mates to a cave or some sheltered place. We imagine our forebears as sharing our domestic ideal of the home as a haven from the elements, the beasts of nature or a heartless world.[3] For most of human history, for most humans, the 'ideal' has remained just that – an unmet potential. We developed increasingly homey homes, adding heat and windows and doors and furnishings, but our time indoors was continually interrupted by tasks that we undertook outdoors, time and activities that sustained meaningful interaction with the flora and fauna; weather and seasons; land, sky and sea. The pace of 'progress' towards realising the ideal of living indoors has sped up dramatically in the last century, so that in North America and to a lesser degree in Europe today, most areas of human activity – whether work, rest, or play – take place in isolation from natural light, 'fresh' air and meaningful interactions with environments outside or beyond the 'built' one, environments that not so long ago dominated our days.[4] Our shelters are amazing temples of technology and comfort, but for well over a century indoor life has been a mixed blessing.

This chapter will focus on some of the problems arising from the 'great migration' indoors and the biotechnological adaptations developed to ameliorate them. It seeks to build upon several robust projects in recent environmental history that deal with human health and the environment, whether focusing on the sites we select for human activity and how (and at what costs to our health) we alter their geography, or issues specific to interior environments, both occupational and domestic.[5] After a brief introduction highlighting one optimistic forecast and some pessimistic assessments from a century ago, the chapter turns to health burdens stemming directly from time spent indoors, exemplified by the deficiency disease rickets. The 'cure' for rickets by biotechnological fixes minimised the need for sunshine, permitting people to spend more time indoors. Increasingly, those indoor spaces are air conditioned, and the third section outlines the psychological and physical stressors arising from habituation to (or aversion to) controlled climates. Air conditioning must be seen in a larger context of the modern built environment and its tendency to promote physical isolation, and the fracturing of traditional socio-spatial relationships – a social outcome, it can be demonstrated, with important health costs. This chapter will argue that even the most successful adaptations to life indoors have important consequences for health, for society and for what it means to be human.

10.1 The move indoors: cheerleaders and nay-sayers

We have long lived with a tension between the love of open air, and the sense of control and comfort that shelter brings. This tension grew into publicly expressed ambivalence in the late nineteenth century as architects, technologists and homebuilders grappled with rapidly developing technologies – from window-screens and balloon framing to ventilation and plumbing to telephone and electricity – all of which held the promise to transform indoor life. Some futurists a century ago embraced the prospect of adopting technology to cradle home dwellers in a technological bubble. In 1907, a *Washington Post* report on two model homes imagined that the typical home a century hence would have '… double walls, vacuum dusters, automatic cooking range, self-operative electric laundry and scullery, front entrance periscope, searchlights, range finder and telephones, winter sun and winter garden, opera, theater and concert room equipped with slot-meter connections with all the best theaters and concert halls …'. This high-tech house would include 'a complete automatic domestic service', be 'perfectly dust-proof' and feature 'all the latest devices for heating, cooking and for ventilation'.[6] These visions were far from fantasy, and the inventors showing off their ideas were not out of mainstream thinking about how technology and encapsulation could improve health and reduce labour.

Not everyone was so sure that indoor life was something to be encouraged. That same year, physician and health promoter John Harvey Kellogg (of Battle Creek sanitarium and Corn Flakes fame) commented on the fate of the tens of thousands of Californians thrown out of their homes by the recent San Francisco earthquake; not surprisingly, he found a health lesson in their ordeal. He speculated: '… if these tent-dwellers get such a taste of the substantial advantages of the out-of-door life that they refuse to return to the old unwholesome conditions of ante-earthquake days, they will profit substantially by their experience'. With gushing sarcasm, he outlined the advantages of indoor life: 'A pale face instead of the brown skin which is natural to his species … an aching head and confused mind and depressed spirits, instead of the vim and snap and energy, mental and physical, and the freedom from pain and pessimism of out-of-door dwellers'. He saw the problems of indoor life in evolutionary terms: 'Man is an out-of-door creature, meant to live amid umbrageous freshness, his skin … browned and disinfected by the sun …' Perhaps we had 'become accustomed to the unhealthful and disease-producing influences of the modern house', but we were not immune to its dangers. 'An atmosphere that will kill a Hottentot or a baboon in six months', he warned 'will also kill a bank president or a trust magnate – sometime'.[7]

Many of the health seekers at Kellogg's sanitarium at Battle Creek were urban or suburban middle-class women, who, having the bulk of their domestic tasks – from shopping to housework to childcare – done by others, and their breadwinning done by their husbands, were the first sizable cohort of Americans with the means and the opportunity to choose, if they so desired, to spend all their days indoors.[8] And it was this indoor life that many blamed for the conditions that brought them to Kellogg and other healers. 'Dark living rooms, offices and workshops are a chief cause of enervation and of physical and mental weaknesses', argued Sydney Dunham in 1896, 'and persons spending their lives in them will always find themselves victims to a great variety of ailments'.[9]

Indoor life quickly became the focus of blame for the neurasthenia and other constitutional infirmities that filled the pages of neurasthenics' diaries and their doctors' casebooks. George M. Beard, the New England neurologist who coined the term neurasthenia, contrasted the carefree outdoor life of the 'Indian Squaw, sitting in front of her wigwam', with the debilitating world of the 'sensitive white woman – pre-eminently the American woman ... living indoors; torn and crossed by happy or unhappy love; subsisting on fiction, journals, receptions; waylaid at all hours by the cruelest of all robbers, worry and ambition'.[10] At Dr John Gehring's institution in Bethel, Maine, members of the professions sought relief from nervous disorders by a regimen of outdoor exercise and recreation.[11]

Concern with indoor life extended beyond the health of individual neurasthenics. Medical and social progressives in North America and Europe found in the modern built environment conditions that endangered the very nature of humanity. Although Edward Carpenter, a British Fabian Socialist, had a larger economic and political agenda, he took the debilitating aspects of this environment quite seriously. His essay, 'Civilisation: Its Cause and Cure', warned of 'a society broken down and prostrate, hardly recognisable as human, amid every form of luxury, poverty and disease'. Modern property relations had turned mankind from nature: 'He who had been the free child of Nature denies his sonship ... He deliberately turns his back upon the light of the sun, and hides himself away in boxes with breathing holes (which he calls houses), living ever more and more in darkness and asphyxia'.[12]

Carpenter cast the results of mankind's choices in (de)-evolutionary terms: 'He ceases to a great extent to use his muscles, his feet become partially degenerate, his teeth wholly, his digestion so enervated that he has to cook his food and make pulps of all his victuals...' The solution, for Carpenter, was a movement towards 'Nature and Savagery'. 'Life indoors and in houses has to become a fraction only, instead of the principal part of existence as it is now. Garments similarly have to be simplified'. In the late 1880s, when he read this essay to the Fabians, he claimed to be optimistic about the prospects of reversing this degrading trajectory, boasting that a 'nature movement' that had started previously in the arts was 'rapidly realizing itself in actual life ... and developing among others into a gospel of salvation by sandals and sunbaths!'[13] Obviously, neither Carpenter's Great Britain nor the United States converted. A critic of Carpenter's proposals 20 years later sniffed: 'The sunbath is only one of the many remedies prescribed to the poor by doctors impaneled by the British state, and the sandals are better made by machinery than by the hands of poetic hermits.'[14]

If the solutions proposed by reformers like Kellogg and Carpenter seemed quirky and quaint a generation later, their dire predictions for human health had put them squarely in the mainstream of public health expertise of their day. The seemingly intractable rates of infant mortality, the crushing burden of tuberculosis, and the abysmal health condition of the poor and working class at the turn of the twentieth century prompted bleak forecasts from all quarters, from advocates for clothing reform and spokesmen for the New Public Health alike.[15] So while the widespread association of indoor air with ill health and race degeneration drove much of the proselytising for 'outside air' via open air schools, sleeping porches and vigorous outdoor exercise, that same association simultaneously provided ongoing impetus for an entirely different solution, one more in keeping with the rapidly industrialising, modern world. Progressive programs of occupational safety, often built upon new studies of human physiology; a burgeoning electrical machinery industry; the steady trend toward larger, taller and more technologically complex buildings: these and other factors fuelled a growing network of technologists, health specialists, and business interests dedicated to controlling *indoor* air quality through technology – through forced-air ventilation and climate control.[16] In short, predictions about the health dangers of too much time spent indoors hastened the advent of the age of artificial climate control.

The irony of this outcome follows a peculiar dialectic that recurred throughout the twentieth century and continues to this day: a population finds its way deeper into indoor life; some problem (whether health-related or sociological) is identified and blame focuses on the surfeit of indoor life; a technical solution is found – one that minimises the particular negative effect of indoor life (as opposed to solutions that actually reverse the inward migration); the solution becomes the new norm, facilitating further 'in-migration'.

10.2   Rickets and its cure by artificial sunlight

Perhaps nowhere is this recurring pattern clearer than in health conditions associated with time spent indoors, and of these health issues, rickets has the most obvious link to indoor life; furthermore, society's response to rickets is archetypal in its whole-cloth recourse to technological rather than natural 'solutions'. Few health experts 50 years ago would have imagined that this disease would ever re-emerge in the industrialised world. But since the turn of the twenty-first century, rickets has staged a small but troubling resurgence.[17] Ironies abound in this development: America, the fattest nation in the world, is grappling with a disease of malnutrition; but it is often cultural practices and health precautions, not lack of food that are responsible. Rickets is a failure of adequate mineralisation in bones due to a lack of vitamin D. In nature, the body can acquire vitamin D from a diet rich in animal fats, or it can produce the vitamin directly, given sufficient exposure to ultraviolet light. In the United States, as in Northern Europe, the greatest risk factors for rickets are dark skin, being breastfed without vitamin D supplementation, and wearing clothing that nearly completely covers all skin. Conscientious parents who shun the products of the commercial food industry (where universal vitamin D supplementation is the norm), or slather their kids with sunscreen (or keep them out of the sun entirely) to reduce the far greater immediate risk of skin cancer, unintentionally put their offspring at risk of the health issues associated with vitamin D deficiency, including rickets, a condition long associated with poverty, malnutrition and bad parenting.[18]

Well before twentieth-century science explained the role of ultraviolet light and diet in the process of vitamin D production and calcium metabolism, doctors and mothers correctly linked rickets to diet and sunless environments. The first published European accounts by Daniel Whistler and Francis Glisson in the seventeenth century focused on diet and too much time indoors.[19] Glisson advocated more light and air to prevent and cure rickets, though more out of adherence to humoral medicine than any understanding of a link between sunlight and bone growth.

As cities grew rapidly in the Industrial Revolution, the changing urban landscape and deteriorating environment drew frequent comment from social critics; meanwhile, physicians reported dramatic increases in rickets. Remarkably, contemporary observers often correctly associated the two trends, as did this late nineteenth-century physician, who observed that rickets arose in settings where children were: 'shut out from the light partly by the height of the houses, partly from the fact that even the sun's rays which do manage to struggle through the canopy of smoke which envelops them, are so diluted that they are of comparatively little value.'[20] Rickets was an enormous problem in Victorian England, and correctly associated with the poor, who were far more likely to fall to respiratory infections aggravated by rickets, and who were least likely to enjoy successful interventions.

American medical professionals began to address rickets in the years after the Civil War, and the particular dynamics of America's demographic and industrial developments shaped both the reality and perception of the disease. As if replicating the British experience, urban growth in nineteenth century America promoted a corresponding growth in rickets, especially among those living in the darkest, most crowded settings. Rickets rose in America during a tumultuous period of animosity between natives and immigrants, a time when racial distinctions played a big role in explaining almost every problem – and 'race' with its particular late nineteenth century flavour, as spelled out in Buffalo physician Irving M. Snow's 1894 study 'An Explanation of the Great Frequency of Rickets among Neapolitan Children in American Cities', which argued 'that the true reason for the very frequent development of the disease in Italians is found in the fact that it is the effect of our cold, damp, northern climate upon the offspring of a race that has lived for many hundred years in the warm, dry, sunny air of Naples and Sicily'. For Snow, rickets demonstrated '...the physical deterioration of a southern race in a northern climate'.[21]

From at least the early nineteenth century in England, rickets was considered a disease of civilisation, an association that persisted well into the twentieth century, both in Europe and North America. 'The disease is found only in civilized communities', *Scientific Monthly* reported in 1925, 'especially in large cities, where children are kept too much indoors, and are fed on a diet deficient in green vegetables and in other sources of vitamin D'.[22] The association was spelled out even more starkly when it concerned dark-skinned people, for whom the prescription often seemed simply to avoid the seats of modern culture. 'City conditions seem to have an unfavorable effect on negro infants', wrote James Tobey in 1926, '... and competent observers believe, that hardly a Negro child in New York escapes rickets in some degree, mild or severe'.[23] 'Among savage peoples', the *Scientific Monthly* continued, 'the children are, in general, free from it, because they are exposed to sunlight every day. The Eskimo baby is free from it, even though he lives in a dark hut, because he is suckled by a mother who consumes great quantities of animal fat and oil'.[24]

Health reformers and researchers responded to the growing epidemic of rickets in northern American cities, focusing on improving diet and increasing children's exposure to sunshine; they pushed for the regular use of the only 'magic bullet' at their disposal – cod-liver oil.

Paediatrician Martha May Eliot sought to test 'whether rickets could be prevented in a community by the intensive use of cod liver oil and sunlight'. Doctors and mothers had employed cod-liver oil for a century, despite science having no adequate explanation for its effectiveness. In Eliot's study, conducted in New Haven from 1923 to 1926, participating mothers received training in how to administer the oil, and brought their children once a month for examinations, including x-rays. Mothers were given detailed instructions regarding sunbaths for their infants. Of the 216 infants who remained in the study's first year, 186, or 86 percent, showed signs of rickets by clinical exam, but comparisons with three control groups gave solid proof that the oil and sunlight reduced the severity of the rickets.[25]

In 1925, the American Public Health Association (APHA) convened a Committee on Nutritional Problems, which concluded that rickets prevention entailed a three-legged approach: a diet sufficient in phosphorus, calcium and 'antirachitic factor' (whether sunlight or vitamin D). The committee advised that pregnant women should have adequate and nutritious diets, with lots of time spent in the open air and sun, and that infants should be placed in the sun every day, and receive daily doses of cod-liver oil. Do this, the APHA promised, and 'rickets would be abolished from the earth'.[26] This three-legged stool never stood, but it makes for a fascinating counterfactual exercise to consider what rickets' fate would have been if cod-liver oil and sunshine remained the only preventives. An earlier public health regime, fixated on dirt and fresh air, might have pushed for a different technological fix that relied on guaranteeing adequate exposure to the sun, as well as foods rich with vitamin D.

Instead, America took a different path. Fortifying dairy and cereal products with vitamin D spared generations the ordeal of choking down spoonfuls of foul-tasting cod-liver oil.[27] Armed with the 'sunshine vitamin', and a painless way to provide it, parents guaranteed that their children could get by without the sun's health-giving rays – a technological fix to allow us to live in the built environment we chose to build, and to spend even more of our time indoors. This so-called universal supplementation was a tremendous public health success – and an ironic one, given that this 'cure' was effected through the 'adulteration' of foods, so soon after the landmark Pure Food and Drugs Act, designed to get adulterants *out* of foods.

Far from overnight success, the supplementation of America's foods was a gradual process, and so too was the elimination of the worst forms of rickets. Rickets was never made a reportable condition, so the data are unreliable and skew toward drastic under-reporting. One attempt in the 1960s to collect nationwide historical data found that annual *deaths* caused by rickets ranged between 160 and 500 during the 1930s, falling significantly in the post-World War II years to the single digits in the late 1950s.[28] But while some concluded that rickets had receded to the level of 'medical curiosity', in the words of a 1966 report in the *American Journal of Public Health,* hospitals in the early 1960s still treated hundreds of rickets cases each year, and the severity of the cases was such that many were treated through their 'Crippled Children's Programs'.[29]

All the same, we developed a myth of vitamin D's instant efficacy, apparent in a 2003 web-based article in *Discovery Magazine* that boasted: 'By 1924 ... children began consuming irradiated milk and bread and, seemingly overnight, the imminent threat of epidemic disease dwindled to a half-forgotten historical event'.[30] The recent uptick in cases of clinical rickets in the medical literature, and growing concerns over potential public health consequences of the widely-reported low levels of vitamin D in the general public suggest universal supplementation was neither universal nor sufficient. It remains to be seen if future solutions will seek to take us into the light of day or simply urge us to increase our dose of 'artificial sunshine'.

Rickets is far from the only health problem directly or indirectly linked to indoor life; the histories of polio, childhood lead poisoning and asthma reveal them to be apt examples as well: polio's fatal twentieth-century transformation followed hard upon decades of hygienic progress that eliminated earlier generations' nearly universal exposure (and subsequent immunity) during infancy and early childhood to the virus; and lead paint is only the most famous of indoor toxicants.[31] Even allowing for cross-national differences and variations across social classes, any number of negative health outcomes are closely associated with the generally sedentary nature of indoor life, including cardiovascular diseases, type-II diabetes and other conditions stemming from obesity. The medical and public health response in almost every case has entailed a biotechnological fix – vaccines for polio; powerful chelators for toxicants; bronchodilators and steroids for asthma; statins for cardiovascular disease; insulin and a growing armamentarium of pancreas-enhancement drugs for type-II diabetes. Our medicine cabinets are growing in scale with our indoor spaces.

10.3   The physical costs of cooling the built environment

Just as important as what we put in our bodies to ameliorate the health effects of indoor living, are the adaptations we make to the physical structures – the buildings and the agglomerations of buildings and streets – into which we put our bodies. 'Man is so adaptable', microbiologist Rene Dubos observed in 1965, 'that he could survive and multiply in underground shelters, even though his regimented subterranean existence left him unaware of the robin's song in the spring, the whirl of dead leaves in the fall, and the moods of the wind – even though indeed all his ethical and aesthetic values should wither'.[32] But we did not need to move into underground shelters for critics to see the red flags. For Dubos, the distance we had already traversed toward that entombment revealed that the pace of change in the built environment far outstripped our bodies' ability to adapt. 'In all countries of Western civilization', he wrote in 1960, 'the largest part of life is now spent in an environment conditioned and often entirely created by technology' so that '...man's contacts with the rest of creation are almost always distorted by artificial means, even though his senses and fundamental perceptions have remained the same since the Stone Age'.[33] More was at risk than mere ethics and aesthetics, though. Man may have 'done away with the need for biological adaptation on his part', Dubos argued, but abandoning the far more dependable strategy of relying on evolutionary adaptation was 'tempting fate'.[34]

Surely, ventilation and climate control are the most significant technologies employed to optimise the built environment to some presumed human bodily norms. And air conditioning is the bête noire for critics of that project. From the beginning of the age of coolants, critics lined up to warn of the debilitating effects of confinement to the narrow range of temperature and humidity modern technology would soon permit. Early in air conditioning's ascendency, no less an environmental determinist than Yale geographer Ellsworth Huntington worried about deepening man's dependence on technology, despite his general approval of artificial climate control. Huntington feared that with the perfection of climate control the human race would 'cease to increase our physiological adaptation to environmental conditions'.[35] A steady (though largely ignored) stream of criticism persisted, from social critics such as Lewis Mumford and Vance Packard, as well as from health experts.[36] Dubos and others continued to worry over air conditioning's potential for diminishing the body's ability to adapt to temperature change. Some modern literature combines the traditional concern with air conditioning's impact on physiological adaptability with new concerns about growing obesity. One recent survey concluded that '... the increased use of central air conditioning and heating may also be contributing to rising obesity levels, because the body expends less energy in temperature ranges associated with climate-controlled settings'.[37] Others focused on what might go wrong when complex central heating, ventilation and air-conditioning (HVAC) systems did not operate as specified, producing overly dry air or, worse, acting as an ideal culture medium and distribution system for infectious disease, irritating or poisonous chemicals, and dangerous mould.[38]

By and large, however, the medical literature treats air conditioning as a prescription for better living. Modern medicine has embraced climate control, from the mid-nineteenth century Florida physician who developed an ice-making machine to cool the rooms of his patients, to the ubiquitous climate control of the modern sprawling medical centre.[39] And though the literature duly acknowledges the problems that result when good HVAC systems go bad, it generally expresses a deep approval of air conditioning for hospitals as well as homes, arguing in fact, that perhaps we should all embrace air conditioning more than we already do.

Air conditioning appears to increase longevity. Studies of coronary death seasonality – the ratio of summer or winter mortality over overall mortality – found that from the 1930s to the 1970s, improvements in home heating reduced winter's particular deadliness. From the 1970s on, however, winter's share in mortality rose again. It was not that winter was growing deadlier once more, but that summer heat was exacting a smaller toll, thanks to increased use of air conditioners.[40] Studies of deaths during heatwaves make the same association. In its report on the 1995 heatwave in the upper Midwest, the *Wisconsin Medical Journal* concluded that 'to reduce the number of illnesses and deaths that occur during future heatwaves, all residents should be advised to install air-conditioning systems in their homes or apartments'; and an analysis of deaths during Chicago's 1999 heatwave found that 'the strongest protective factor was a working air conditioner'.[41] Outside the United States, air conditioning's benefits during heatwaves were also applauded: one study of mortality in France's fatal 2003 heatwave did not mention air conditioning in private residences, but found that hospitals whose intensive care units were air conditioned fared better in preventing heat-related deaths.[42]

If air conditioning is making us healthier, it is not because Dubos and Huntington were wrong; Americans' bodies are not becoming more heat tolerant. On the contrary, public health and medicine have embraced climate control because it accommodates the growing culture-bound *intolerance* of seasonal extremes, a consequence of our desire no longer to be 'warmed and cooled by the same winter and summer' as were our grandparents.[43] Modern Americans have, in effect, developed an artificially cooled exoskeleton comprising their homes, their automobiles and most of the other places in which they spend their days. Should we applaud the resulting longevity and greater years of vitality, and reduced exposure to the risks of outdoor living? Or should we worry – and even complain – that we may be jettisoning long-lived human abilities as we turn our care over to the machines? We should at least consider whether comparable health benefits could accrue from making a different set of changes to the built environment. After all, our homes, air conditioned or not, exist in a setting largely defined by social factors: the facts of their physical construction; their site; and their location relative to the buildings to which we travel for work, leisure and other needs. Such physical and social factors play a large role in health, as reports on the causes of heatwave-related mortality make clear: elderly people living on top floors are more isolated, hence at greater risk of missing out on such simple social preventives as having neighbours check in to make sure they are drinking water and comfortable; in neighbourhoods with high perceived crime rates, social interaction drops and windows stay shut.[44] The healthfulness and desirability of life indoors depends on any number of infrastructural factors, many of which are amenable to change, and which are the focus of public health initiatives involving sprawl reduction, energy efficiency, pollution and the economics of food distribution.

10.4 Isolation and new networks: the implications for health

Life indoors is comfortable but socially isolating, as Dubos observed in 1968: 'In many respects, modern man is like a wild animal spending its life in a zoo; like the animal, he is fed abundantly and protected from inclemencies but deprived of the natural stimuli essential for many functions of his body and his mind. Man is alienated not only from other men, not only from nature, but more importantly from the deepest layers of his fundamental self'.[45] This isolation has social and political costs, many of which impact human health. For over 30 years, public health studies exploring the connection between 'social capital' and health have tested the positive association between adequate social networks and health. The predictable negative impact of sequestration in heatwave morbidity and mortality is one of the most commonly cited examples; social support is positively correlated with healthy child development, improved mental health, stress reduction, prevention of environmental health risks and participation in health enhancement programs.[46]

If technology in the form of artificial climate control has driven us apart, perhaps the technology of telecommunications – both active and passive – can reduce some of the resulting social isolation. Good or bad, telecommunications have redefined social interaction several times over. Robert and Helen Lynd noted the mixed blessing of radio in the 1920s, which drew folks off the semi-public front porch and into the privacy of their home's interior, while simultaneously serving as a focus for shared family time and bringing the wider world to 'Middletown's' citizens.[47] Television, of course, had a veritable army of statisticians and curmudgeons watching its every tic and growing pain, with no small number of them noting the medium's tendency to keep families at home. In 2000, Robert Putnam assigned a good portion of blame on television's isolating tendency for the loss of social capital in America.[48]

But though Putnam worried about moderns bowling alone, at least his solo bowlers were out of the house at the bowling alley. In the home of the future, communications technologies may make even that trip unnecessary. Already, many retirement homes and senior centres provide virtual bowling via the wireless videogame system, the Wii. In an ironic twist, the popularity of this 'sport' has given rise to hundreds of bowling teams in the United States competing in a nationally organised competition for seniors.[49] As for the internet, which Putnam mused had not yet proven whether it would become 'a really nifty telephone or a really nifty television', the answer would appear to be 'both'.[50] Social networking falls squarely on the 'nifty telephone' side, and is one of the internet's most promising avenues for rebuilding social capital. So, as impressive as geriatric Wii bowlers may be, a more fitting vision for telecommunications' impact may be the obese young adult reading and posting to Facebook while streaming the latest episode of 'The Biggest Loser'.

One of the few bright spots in this bleak terrain is wireless communications, from cell phones to wi-fi access to the internet, which mark one of the few technologies that make spending more everyday time outdoors an attractive option. But bringing more people back out into the commons does not necessarily bring them together. Nearly everyone out there on the green, it seems, is yakking into their cell phones, text messaging or consuming increasingly balkanised popular media. Whether these new media produce an upswing in social capital is one question.[51] Whether any health benefits accrue from this evolving geographically fractured networking is dubious.

10.5 The gardener in the machine

A different sort of evolution, of course, is at the heart of this essay, which so far has avoided the conundrum declared in its title. In evolutionary terms, *homo sapiens* is much closer to the hunter-gatherer and gardener than to today's web-surfing, information-harvesting couch potato. This suggests an evolutionary crisis with which the species is only beginning to grapple. The central dynamic I have described can fruitfully be seen as some form of evolution, in that it is changing what it means to be a human. We are living through a gradual, deep, and widely-shared transformation in our relationships with our bodies and our environments – both the elemental, 'natural' aspects, but especially the environments in which we surround ourselves.

It is an evolutionary surge of unprecedented speedy onset: since Neolithic humans took up farming and began domesticating animals, they started down a long path of co-evolution with the germs and parasites of their beasts, and the microbes stirred up by their agriculture. It took us over ten thousand years to become gardeners, and gardeners we remain. Our subsequent move to the indoors, rising out of the muck and manure and the pollen and the sun, has been practically instantaneous, and we have no natural mechanisms to change that quickly. It is a frightful prospect. Our images of man's future, from H.G. Wells to *Star Trek*, feature Lamarckian nightmares of Morlocks and Eloi; of long-fingered, big-headed weaklings; or effeminate beautiful people pampered by – or worse – their machine servants. Criticism of indoor life is frequently couched in a rhetoric of degeneration, so pithily captured in Carpenter's description of the unrecognisable crypt-dwellers we might become.

And yet the comfort and control of the great indoors continues to beckon, to draw us further from the 'umbrageous freshness' Kellogg declared as our birthright. And so we are doing the only thing we can, wriggling like the hermit crab into our ever-improving biotechnological exoskeletons, slouching along a new evolutionary road toward what some call post-humanity.[52] Most of the discourse about post-humanity – whether enamoured of it or repulsed by it – focuses on the most intimately embodied biotechnologies, from prosthetics and biorobotic hybrids, to nanotechnology to genetic engineering – in short, embodied enhancements to body function. Generally the criticism is framed largely as the futurist's bailiwick – concerns about a slippery slope we might step upon.

The history outlined in this chapter suggests that to the degree that our built environment and biotechnological adjuvants serve to extend (or obviate) our bodies' capacity for adaptation, we have already been on that slippery slope a long time, that we have long been engaged in a form of evolution – if not Darwinian or Mendelian, then social and technological. We are undergoing a process of biotechnological Lamarckism: our technological choices become acquired traits humans pass to our offspring through that human adaptation called culture. Since this evolution is social, and engineered, our ideas and decisions are critically important.

Early on, we decided the ideal of the home as a haven was a good one, worthy of support. And so we developed a number of technologies to make life indoors more comfortable to the normed notion of the body (and the body politic). These are technologies that make us seem 'super-natural' – not in the comic book superhero sense, nor the spoon-bending or clairvoyant – but in the sense that we have risen above nature's evolutionary baggage. Sigmund Freud's term for the general notion was that we had 'become a kind of prosthetic God'.[53] But gods, one would hope, are not routinely poisoned by, debilitated by, and enslaved by their prostheses.

Notes

1. T. C. Rick and J. M. Erlandson (2009). 'Perspectives: Anthropology: Coastal exploitation', *Science*, 325, 5943, pp. 952–3.
2. R. Dubos (1968). *So Human an Animal: How We are Shaped by Surroundings and Events*, orig. publ. New York: Charles Scribner's Sons; repr. ed. (1998) with a new introduction by J. Cooper and D. Mechanic (New Brunswick, New Jersey: Transaction Publishers).
3. J. C. Berman (1999). 'Bad hair days in the paleolithic: Modern (re)constructions of the cave man', *American Anthropologist*, new series 101, pp. 288–304.
4. J. M. Samet and J. D. Spengler (2003). 'Indoor environments and health: Moving Into the 21st century', *American Journal of Public Health*, 93, pp. 1489–93.
5. L. Nash (2006). *Inescapable Ecologies: A History of Environment, Disease, and Knowledge* (Berkeley: University of California Press); C. Bolton Valencius (2002). *The Health of the Country: How Americans Understood Themselves and Their Land* (New York: Basic Books); G. Mitman, M. Murphy and C. Sellers (eds) (2004). 'Landscapes of exposure: Knowledge and illness in modern environments', *Osiris*, 19, pp. 1–304.
6. Anon. (1907). 'Home a hundred years hence', *The Washington Post*, 27 October, p. M2.
7. G. Wharton James (1908). *What the White Race May Learn from the Indian* (Chicago: Forbes & Co.), p. 65.
8. P. Gestner (1996). 'The temple of health: A pictorial history of the Battle Creek Sanitarium', *Caduceus: A Humanities Journal for Medicine & the Health Sciences*, 12, 2, pp. 1–99.

9. S. A. Dunham (1896). 'The air we breathe', *The Chautauquan: A Weekly Newsmagazine*, 23, 3, p. 279.

10. G. M. Beard (1884). *Sexual Neurasthenia: Its Hygiene, Causes, Symptoms and Treatments*, 6th edn, ed. Alphonse David Rockwell (New York: E.B. Treat & Co., 1905), pp. 59–60.

11. D. G. Schuster (2005). 'Personalizing illness and modernity: S. Weir Mitchell, literary women, and neurasthenia, 1870–1914', *Bulletin of the History of Medicine*, 79, pp. 695–722; B. Harris and C. Stevens (2010). 'From rest cure to work cure', *Monitor on Psychology*, 41, 5, p. 26.

12. E. Carpenter (1916). *Civilization: Its Cause and Cure, and Other Essays*, 14th edn (London: George Allen & Unwin, Ltd.), pp. 26–7.

13. Ibid., p. 49.

14. S. Coit (1917). *Is Civilization a Disease?* (Boston and New York: Houghton Mifflin), p. 130.

15. D. Porter (1991). '"Enemies of the race": Biologism, environmentalism, and public health in Edwardian England', *Victorian Studies*, 34, 2, pp. 159–78.

16. G. Cooper (1998). *Air Conditioning America: Engineers and the Controlled Environment, 1900–1960* (Baltimore, MD: Johns Hopkins University Press).

17. S. H. S. Pearce and T. D. Cheetham (2010). 'Diagnosis and management of vitamin D deficiency', *British Medical Journal*, 340: b5664; J. Hope, 'The Return of rickets: Victorian disease on the rise due to poor diet and lack of exercise, *Mail Online*, http://www.dailymail.co.uk/health/article-1244988, accessed 23 May 2010.

18. Centers for Disease Control, 'Skin cancer statistics', http://www.cdc.gov/cancer/skin/statistics/index.htm, accessed 19 July 2010.

19. P. Dunn (1998). 'Francis Glisson (1597–1677) and the "discovery" of rickets', *Archives of Disease in Childhood Fetal and Neonatal Edition*, 78, 2, F154-5.

20. W. Macewan (1880). *Osteotomy, With an Inquiry Into the Ætiology and Pathology of Knock-Knee, Bow- Leg, and Other Osseous Deformities of the Lower Limbs* (London: Churchill) quoted in D. Gibbs (1994). 'Rickets and the crippled child: An historical perspective', *Journal of the Royal Society of Medicine*, 87, pp. 729–32.

21. I. M. Snow (1894). 'An explanation of the great frequency of rickets among neapolitan children in American cities', *Transactions of the American Pediatric Society*, 6, pp. 160–76.

22. W. Craig and M. Belkin (1925). 'The prevention and cure of rickets', *Scientific Monthly*, 20, 5, pp. 541–50.

23. J. O Tobey (1926). 'The death rate among American negroes', *Current History*, 25, pp. 217–9.

24. Craig and Belkin, 'Prevention and cure of rickets', p. 542.

25. M. M. Eliot (1925). 'The control of rickets: Preliminary discussion of the demonstration in New Haven', *JAMA*, 85, pp. 656–63.

26. Anon. (1926). 'The prevention of rickets', *AJPH*, 16, 2, pp. 139–41.

27. R. Apple (1989). 'Patenting university research: Harry Steenbock and the Wisconsin Alumni Research Foundation', *Isis*, 80, 3, pp. 374–94.

28. M. T. Weik (1967). 'A history of rickets in the United States', *American Journal of Clinical Nutrition*, 30, pp. 1234–41.

29. H. E. Harrison (1966). 'The disappearance of rickets', *AJPH*, 56, pp. 734–7.

30. R. Conlan and E. Sherman (2003). 'Unraveling the enigma of vitamin D', http://www.beyonddiscovery.org/content/view.txt.asp?a=414#Tracing_the_Cause_of_Disease (National Academy of Sciences, accessed 15 May 2010).
31. N. Rogers (1992). *Dirt and Disease: Polio Before FDR* (New Brunswick, New Jersey: Rutgers University Press); C. Warren (2000). *Brush With Death: A Social History of Lead Poisoning,* (Baltimore, MD: Johns Hopkins University Press).
32. R. Dubos (1965). *Man Adapting* (New Haven: Yale University Press), p. 279.
33. Dubos, *So Human an Animal*, p. 16.
34. R. Dubos (1959). *Mirage of Health: Utopias, Progress, and Biological Change* (New York: Harper & Brothers), pp. 46–7.
35. M. E. Ackermann (2002). *Cool Comfort: America's Romance With Air-Conditioning,* (Washington, DC: Smithsonian Institution Press), p. 144.
36. Ibid., pp. 146–9.
37. D. E. Jacobs, J. Wilson, S. L. Dixon, J. Smith and A. Evens (2009). 'The relationship of housing and population health: A 30-year retrospective analysis', *Environmental Health Perspectives*, 117, 4, pp. 597–604.
38. H. T. Smedbold, C. Ahlen, S. Unimed, A. Nilsen, D. Norbaeck and B. Hilt (2002). 'Relationships between indoor environments and nasal inflammation in nursing personnel', *Archives Of Environmental Health*, 57, 2, pp. 155–61; V. L. Yu (2008). 'Cooling towers and legionellosis: A conundrum with proposed solutions', *International Journal of Hygiene And Environmental Health*, 211, pp. 229–34.
39. R. Arsenault (1984). 'The end of the long hot summer: The air conditioner and southern culture', *The Journal of Southern History*, 50, pp. 597–628.
40. D. Seretakis et al. (1997). 'Changing seasonality of mortality from coronary heart disease', *JAMA* 278, pp. 1012–14.
41. L. Knobeloch, H. Anderson, J. Morgan and R. Nashold (1997). 'Heat-related illness and death, Wisconsin, 1995', *Wisconsin Medical Journal*, 96, 5, pp. 33–8; M. P. Naughton, A. Henderson, M. C. Mirabelli et al. (2002). 'Heat-related mortality during a 1999 heat wave in Chicago', *American Journal of Preventive Medicine*, 22, 4, pp. 221–7.
42. B. Misset, B. DeJonghe, S. Bastuji-Garin et al. (2006). 'Mortality of patients with heatstroke admitted to intensive care units during the 2003 heat wave in France: A national multiple-center risk-factor study', *Critical Care Medicine*, 34, 4, pp. 1087–92.
43. R. S. Kovats and S Hajat (2008). 'Heat stress and public health: A critical review', *Annual Review of Public Health*, 29, pp. 41–55.
44. E. Klinenberg (2002). *Heat Wave: A Social Autopsy of Disaster in Chicago* (Chicago: University of Chicago Press).
45. Dubos, *So Human an Animal*, p. 16.
46. L. F. Berkman and S. L Syme (1979). 'Social networks, host resistance, and mortality: A nine-year follow-up study of Alameda County residents', *American Journal of Epidemiology*, 109, pp. 186–204; S. Szreter and M. Woolcock (2004). 'Health by association? Social capital, social theory, and the political economy of public health', *International Journal of Epidemiology*, 33, pp. 650–67.

47. R. S. Lynd and H. Merrell Lynd (1929). *Middletown: A Study in Modern American Culture* (New York, Harcourt, Brace and Company, repr. ed. 1956 Harvest Books), p. 269.
48. R. Putnam (2000). *Bowling Alone: The Collapse and Revival of American Community* (New York: Simon & Schuster).
49. J. Brockman (2009). 'Who's gaming now? Seniors turn to Wii bowling', *All Things Considered*, 23 November, archived at http://www.npr.org/templates/story/story.php?storyId=120705467, accessed 13 December 2009.
50. Anon. (2000). 'The Other Pin Drops', *Inc.*, 15 May, archived at http://www.inc.com/magazine/20000515/18987.html, accessed 23 May 2010.
51. A. Quan-Haase and B. Wellman (2004). 'How does the Internet affect social capital?' in Marleen Huysman and Volker Wulf (eds). *Social Capital and Information Technology* (Cambridge, MA: MIT University Press), pp. 113–31.
52. R. Kurzweil (2005). *The Singularity is Near: When Humans Transcend Biology* (New York: Viking); N. Katherine Hayles (1999). *How We Became Posthuman: Virtual Bodies in Cybernetics, Literature, and Informatics* (Chicago: University of Chicago Press); F. Fukuyama (2002). *Our Posthuman Future: Consequences of the Biotechnology Revolution* (New York: Farrar, Straus and Giroux).
53. S. Freud (1930). *Civilization and its Discontents*, trans. and ed. James Strachey (New York: Norton, 1962), pp. 38–9.

11
Exposing the Cold War Legacy: The Activist Work of Physicians for Social Responsibility and International Physicians for the Prevention of Nuclear War, 1986 and 1992

Lisa Rumiel

When the Chernobyl nuclear reactor meltdown sent gusts of radioactively contaminated air across the globe in 1986, both International Physicians for the Prevention of Nuclear War (IPPNW) and its American affiliate, Physicians for Social Responsibility (PSR), set about their antinuclear activism with renewed vigour. PSR alone was swamped with phone calls from people living near nuclear reactors, weapons-testing sites, manufacturing facilities, waste dumps and uranium mines, as well as from Americans travelling and living abroad. People expressed concerns about the long and short-term effects of radiation exposure; they were anxious about the ecological threats of radiation; they wanted information on the difference between a commercial nuclear reactor meltdown and the fallout from a nuclear weapon explosion; and they worried about 'the psychological aspects of nuclear crisis management'.[1] Just six days after the accident, PSR orchestrated its largest press conference ever through its national office. The group called for American-Soviet negotiations to set up an 'international protocol for cooperative management of disasters involving nuclear technology', advocated the establishment of an international panel of scientists to study the long and short-term health effects of the Chernobyl disaster, and encouraged the immediate shut-down of all Department of Energy (DOE) 'operated reactors until they [could] meet Nuclear Regulatory Commission (NRC) standards'.[2] There were over 20 major press outlets and nine cameras in attendance to hear PSR's reaction to Chernobyl, while the *New York Times* and the *Washington Post* phoned after the briefing, requesting a summary and a press packet.[3] PSR spokesperson, Jack Geiger, also appeared on the television shows, *CBS Morning* and *CBS Nightly News*, the *Larry King Show*, and *Nightline* in the week after the meltdown, talking about the long and short-term effects of the accident and warning that 'the worst is yet to come'.[4]

The media attention showered on IPPNW in the aftermath of the accident was even greater, due mainly to that fact that the group was still riding high from its 1985 Nobel Peace Prize award.[5] There were over 500 media outlets in attendance at its sixth World Congress, which was held in Cologne, West Germany six weeks after news of the Chernobyl accident reached the international community. The highlight of the World Congress was the speech given by the Russian delegates, Yevgeny Chazov (the Soviet president of IPPNW) and Leonid Ilyin (a Soviet surgeon and chairman of the USSR Committee on Radiation Protection), who were both part of the post-Chernobyl emergency response medical team. They talked about the medical effects of the Chernobyl explosion, the medical response to the disaster, and also the comparison of Chernobyl to a nuclear bomb explosion. The public health impact and the complete devastation caused by the accident were used to reinforce the core message shared by IPPNW and PSR – there could be no effective medical response to nuclear war. According to the Russian physicians, Chernobyl killed 29 people; yet, the entire Soviet medical system was mobilised to deal with the accompanied devastation. This included the assembly of 230 medical teams, with a total of 5000 doctors and nurses who saw to all of the 100,000 people evacuated from the area. Chazov and Ilyin underlined that the same would not be possible with a larger accident. The *New York Times* reported Chazov's words, 'medicine will be helpless if even a few nuclear bombs are detonated'.[6] The fact that the Soviet physicians were part of the disaster relief after Chernobyl not only bolstered the group's claims of medical expertise; it also strengthened its declaration of authority on the need for nuclear disarmament.[7]

This article offers an analysis of PSR and IPPNW's anti-nuclear activism between 1986 and 1992, beginning immediately after the Chernobyl accident on 26 April 1986. My interest is in building an understanding of how the physicians' movement (first in the US and later on the international stage) finally came to recognise and raise awareness about the links between militarism, environmental contamination and human health by the late 1980s. Since each groups' establishment (PSR in 1979 and IPPNW in 1981), their focus had been on increasing public awareness about the medical consequences of nuclear war. After 1982 each group was singularly devoted to the Comprehensive Test Ban Treaty (CTB) campaign. The shift towards expressing concern about the immediate health and environmental consequences of the nuclear arms race took PSR and IPPNW members out of their comfort zones and required that the physicians' movement develop a new approach to activism. It was no longer enough to speak in hypothetical terms about the medical consequences of nuclear war. For both groups this change in focus was dependent on several external factors; the Chernobyl accident is only one of them. Indeed, earlier attempts by PSR members to raise awareness about the public health problems of commercial nuclear reactors failed because physicians were ill-equipped to tackle the complex scientific questions surrounding radioactive contamination and the ways it affected both the environment and human health in the long term. While both groups were quite successful in making a meaningful contribution to activism in this area after 1986, this was only made possible through coalition building with like-minded activist groups of physicists, epidemiologists, nuclear engineers and other scientists with relevant expertise for illustrating the relationship between the arms race, environmental contamination and human health.

11.1   The history of the physicians movement

While PSR's first incarnation actually coincided with Rachel Carson's publication of *Silent Spring* in 1962, there is not much to suggest that the group saw itself as part of this new environmental movement. Rather, PSR was part of a growing cadre of expert activists who formed non-profit organisations in the post-World War II period because of its concerns with US military policy, in particular, the nation's expanding capabilities for nuclear destruction. One of the first announcements of PSR's formation, for example, explained that its purpose as a group of 'intelligent and scientifically oriented' minds was to prevail over those in the US military who were 'obsessed with the technics of destruction'.[8] These physicians were clearly influenced by the peace movement and saw a unique role for doctors within that movement. Much of PSR's early work was devoted to exposing the fallacy of American civil defence schemes, arguing that 'there [wa]s no rational basis for such plans'.[9] Its unique contribution to the peace movement was the group's ability to frame its criticisms of nuclear aggression in terms of the terrible health costs of such a war. It was from this argument that PSR developed its classic analysis, 'The Medical Consequences of Thermonuclear War', first published in 1962 in the *New England Journal of Medicine* (NEJM).[10] The many symposia held on this topic between 1962 and 1970, when the group disbanded, were preoccupied with predicting the death toll from a nuclear attack and the total breakdown of medical infrastructure.

The links between PSR (as it was reincarnated in 1979) and IPPNW were rooted in PSR's early history as well as the intensification of the arms race during the early 1980s. Several of the original PSR members, namely Bernard Lown, Geiger, Richard Feinbloom and Victor Sidel, joined the new PSR in 1979 or joined IPPNW's leadership in 1981. For instance, PSR was revitalised by old and new physicians, including Helen and Bill Caldicott, Ira Helfland, Eric Chivean, Feinbloom (from the earlier movement), Rick Ingrasi, Katherine Kahn and Andy Kramer.[11] While these were the physicians to join the first few meetings, people such as Geiger quickly rejoined the group and played a leadership role for many years.[12] On the other hand, IPPNW was Lown's baby. Along with American physicians James Muller and Chivean (one of PSR's founders in 1979), Lown travelled to Geneva in 1980 to discuss the creation of an international doctors' movement with Chazov, Ilyin and Mikhail Kuzin from the Soviet Union. The result of this meeting was the establishment of IPPNW, which became the international umbrella organisation for different doctors' organisations around the world. In 1982, the PSR Board of Directors voted to join IPPNW and become the official American affiliate.[13] Being part of an international alliance, particularly one that was founded by American and Soviet physicians, was a powerful illustration of the possibility for peace between the superpowers and an opportunity to increase exposure for physicians' groups that were concerned about the escalation of the arms race.

Like PSR in its first incarnation, both groups encouraged peace negotiations between the superpowers and their medical assessments of nuclear war were offered up to demonstrate the urgent need for such a peace. Likewise, PSR continued to hold conferences and symposia on the 'Medical Consequences of Nuclear War' in all major American cities, while IPPNW used its yearly congresses to speak about these issues.[14] Each physician speaker took great pains to root his or her discussions of the medical consequences of nuclear war in the scientific method; however, most of these talks were recycled from the original PSR article that appeared in the *NEJM* in 1962 and were not significantly shaped by the particular expertise of physician members. The most popular topic among PSR representatives was the medical effects of a thermonuclear weapon being dropped on a major metropolitan centre in the US. Geiger's discussion of a one-megaton bomb being dropped on Detroit in the IPPNW published volume, *Last Aid,* is a good illustration of this kind of speech. He used a diagram of the city centre, which was broken down into six concentric circles, 'each demarcating a zone of destruction in which the magnitudes of radiation, blast, and heat, and their effects on buildings, can be roughly estimated'. Using this model, Geiger made a series of predictions related to the immediate and long-term devastation that would be caused by such a blast.[15] While it was impossible for either PSR or IPPNW to divorce these discussions from the recognition that nuclear war also accompanied severe environmental contamination, very little serious consideration was given to this issue in any of their public lectures or published articles prior to 1986.[16]

By and large, neither PSR nor IPPNW members possessed the methodological tools to address the relationship between the environment and human health. Helen Caldicott was a paediatrician with expertise in the treatment of children with cystic fibrosis, Lown was a renowned cardiologist at the Harvard School of Medicine, Chazov was at one time the Soviet Minister of Health,[17] and the majority of PSR and IPPNW members came from a variety of other medical specialties focused mainly on the treatment of diseases and ill health in people. This was one of the main reasons that, after PSR briefly joined the international opposition to commercial nuclear power in 1979, both PSR and IPPNW steered clear of the issue after 1980. Helen Caldicott's 1977 testimony in opposition to the construction of another nuclear reactor at the Pilgrim Nuclear Reactor Generating Station in Massachusetts is a good illustration of how the absence of very specific expertise excluded PSR members from the debate between 'credible' scientists in this area. Caldicott's testimony, while certainly informed by the relevant studies that raised concerns with the health effects of low-level radiation exposure, was anecdotal and not based on any of her own research expertise. Because of this, the attorney for the utility company forcefully undermined her testimony by repeatedly highlighting for the Nuclear Regulatory Commission her lack of expertise.[18] Since emphasising medical expertise was so central to how PSR and IPPNW positioned themselves as part of the expert activist wing of the anti-nuclear movement, it made practical sense to avoid commenting on the public health effects of commercial nuclear power after 1980.

Ironically, both groups continued to avoid commenting directly on commercial nuclear power even after Chernobyl, which happened at a commercial reactor and inspired the resurgence of activism in both organisations. In the official PSR press release after the accident, the second paragraph underlined that the group took 'no organizational position on the issue of nuclear power'.[19] Regardless, however, there was considerable overlap between the issues the doctors' movement avoided in the early 1980s and the ones being raised by PSR directly after the Chernobyl accident, followed by IPPNW at the end of 1988. Indeed, similar health and environmental concerns existed for all forms of nuclear technology at all stages along the nuclear fuel cycle and the issue of nuclear waste disposal was universally pressing. PSR and IPPNW's new focus on the relationship between environmental contamination and human health is so interesting because of the groups' earlier efforts to create a distance between their work and the opposition to commercial nuclear power, which had always confronted these issues head on. The fact that both groups tried to maintain this barrier even after Chernobyl is bewildering.

11.2   Environmental health: trends in twentieth-century medicine

The inability of PSR and IPPNW physicians to engage in scientific discussions about the subtle relationship between radioactive contamination of the environment and human health is reflective of broader trends in twentieth-century medicine. Linda Nash aptly observes, 'contemporary medicine does not much concern itself with the landscape. Physicians generally confine themselves to the terrain of the human body, while the natural environment is left to a host of other disciplines'.[20] Two notable exceptions to this twentieth century trend were physicians who worked in Industrial Hygiene or Radiation Health and Safety.[21] However, both of these fields were marginal in the biomedical health sciences and relied heavily upon inter-disciplinary collaboration with relevant technological and scientific disciplines. Moreover, as more technical and complex scientific knowledge about chemical and radioactive toxicity developed, the role of physicians within these two fields became more secondary. This was especially true for the post-World War II era, when increasing fragmentation and specialisation within the sciences intensified the alienation of physicians from developing advanced understandings of environmental health issues.[22] Soraya Boudia's examination of the International Commission on Radiological Protection (1CRP) highlights this trend within the field of Radiation Health and Safety: after 1950 there was not one physician among the 12 commission members or on the six sub-committees, whereas when the ICRP was formed in 1928, the number of physicist and physician commissioners was almost equivalent.[23] My own research into the public debates over the expansion of commercial nuclear power in the US during the 1970s reveals that both the Atomic Energy Commission (AEC) and the NRC favoured the scientific contributions of physicists and engineers for establishing public health and safety policies, regardless of the fact that they had no training in medicine or public health.[24]

None of PSR or IPPNW's leadership was culled from the exclusive group of biomedical researchers who worked on setting radiation protection standards during the Manhattan Project or on any of the various regulatory commissions that were established in the latter half of the twentieth century. Likewise, the prominent place of radiology and nuclear medicine, two medical specialties with vested interests in the continued use of radiation technology for biomedical research, diagnostics, and the treatment of diseases like cancer, also made it less likely that physicians would actively engage in environmental and public health critiques of radiation technology. Indeed, Katherine Zwicker, whose PhD thesis is about biomedical research during the Manhattan Project, reveals that after the war Manhattan Project physicians who were previously focused on radiation health and safety research successfully petitioned for AEC funding so they could resume doing the cancer research that many of them had been doing prior to the war. Thereafter, few of them continued doing health and safety research that dealt with the intersections of environment and health.[25] According to J. Samuel Walker, there was almost no regulation of medical uses of radiation in the 1970s and many physicians sought to keep it that way. For example, when the AEC and the NRC sought to 'strike a balance between' their efforts to regulate radiation exposure and that of the medical licensees in the US, the agencies met with resistance from both individual physicians and professional medical associations.[26]

11.3   Chernobyl and the shifting winds of the doctors' movement

So, why, then, did PSR and IPPNW so enthusiastically embrace the intersections between the radioactive contamination of the environment and human health in the aftermath of Chernobyl? It was not only Chernobyl that influenced this shift. In the American context, Terrence R. Fehner and F. G. Gosling observe that there were important changes in environmental law, revelations about the mismanagement of environmental health and safety within the DOE-run nuclear weapons complex, and increasingly sophisticated environmental actions staged across the US in the three years before Chernobyl. For example, two environmental groups successfully filed a legal challenge against the DOE, citing the violation of the 1976 Resource Conservation and Recovery Act. Despite the DOE's attempt to hide behind provisions within the 1954 Atomic Energy Act, which prohibited states from holding nuclear weapons complex facilities accountable for improper disposal of hazardous wastes, the courts found in favour of the environmental groups in the historic 1984 *Leaf v. Hodel* decision. This decision essentially limited the DOE's ability to self-regulate the nuclear weapons complex's environmental, health and safety programs.[27] Two months before Chernobyl, the DOE was also ordered to release 'approximately nineteen thousand pages of reports' from the Hanford Engineer Works under the Freedom of Information Act. The 'newly declassified' reports revealed a long history of environmental pollution and deception.[28] Newspapers in the Northwest like the *Tri-City Herald, Spokane-Review, Spokane Chronicle, Seattle Post-Intelligencer, Seattle Times*, and *Portland Oregonian* were immediately consumed by this story, but the one *New York Times* reference to the story was buried on page 35.[29] The hysteria caused by the immediate devastation of Chernobyl changed this, making the domestic problem of environmental contamination at places like Hanford front page news across the country.[30] The files also stimulated further investigation into other nuclear weapons facilities, several inquiries at the Congressional level, and inspired legal action at all of the nuclear weapons complex facilities.[31]

PSR's new phase of activism, especially, must be considered within this context. The group's call to shut-down all DOE nuclear reactors in its first press release after Chernobyl guided the focus of its activism for several years after.[32] Because DOE reactors were not regulated by the NRC and many of them were without containment structures around the reactor core, which is what contributed to the Chernobyl meltdown, PSR devoted itself to raising awareness about the shortcomings of these reactor designs. The fact that the entire nation, including several Congressional and Senate Committees, the Justice Department, the NAS and the DOE leadership acknowledged the immense environmental and public health problems caused by the nuclear weapons complex gave the group the confidence to embark on this next phase of work. The group's new focus came at a crucial time in the American anti-nuclear movement. Since Ronald Reagan's re-election in 1984, the issue of disarmament had been losing popularity in the media and among the American public, as is signified by the end of a nationally mobilised Freeze Movement around the same time.[33] In 1986, PSR seized the opportunity to re-captivate the American public and the media with its new message.

This required that the group significantly refine its activist strategies. It was no longer sufficient to rally a group of random medical specialists – for instance, cardiologists, paediatricians, general practitioners, and the like – under the banner of concern for human life. In order to be taken seriously in this new phase of the anti-nuclear debate, the group needed to align itself with relevant experts who could challenge the opposition. Most of the 'experts' serving as spokespersons for the media in the aftermath of Chernobyl had backgrounds in public health, community health, international health, or radiology. Along with Geiger, who was a professor of community medicine with a specialty in medical epidemiology, Jennifer Leaning, a public health specialist who focused on international health issues, Herbert Abrams, who was a radiologist, Sidel and Anthony Robbins, both of whom were also public health experts and who served as presidents of the American Public Health Association, acted as PSR spokespersons after the accident.[34] The presence of these physicians at the press briefing gave the group clout and established these particular representatives as potential 'experts' for the press to rely on when trying to understand the complicated issues involved in a Chernobyl-like accident.

The national PSR offices also relied heavily on a few key PSR chapters across the country to drive home its call to shut down aging and unsafe DOE reactors. Whereas those physicians at the centre of the disarmament debate were mainly situated in the North-Eastern States, PSR chapters in Washington and Colorado State became increasingly important to the overall work of the organisation after Chernobyl. For example, PSR Seattle held a press conference and received widespread media coverage after the Chernobyl accident when it called for the shutdown of Hanford. This chapter also had two physicians from the Hanford area travel to Washington DC to testify at a 'House Subcommittee hearing on nuclear power plants and the danger that the Hanford power plant presents to surrounding communities'. The PSR Chapter in Denver, Colorado, which was located 15 miles southwest of Rocky Flats, was also 'swamped with calls' after news of Chernobyl spread.[35] PSR used this as an opportunity to expand its support in these areas, while also highlighting the similar concerns Americans should have about nuclear safety at home. The importance of regional chapters eventually spread to those surrounding the other key weapons complex facilities, notably, the Nevada test site, the Savannah River plant, and the Fernald plant.[36]

By late 1987, PSR also joined a coalition with eight other scientific and environmental groups to launch the Plutonium Challenge. For a group like PSR, this coalition with engineers, physicists, ecologists, biologists and other environmental/nuclear scientists was crucial for increasing its credibility in this new debate over the environmental and public health threats of the nuclear weapons complex. This also reinforces the observations of Christopher Sellers, who emphasizes that the development of biomedical expertise in environmental health issues is contingent on interdisciplinary cooperation between physicians and scientists.[37] Together, the Environmental Policy Institute, the Energy Research Foundation, the Federation of American Scientists, Friends of the Earth, Greenpeace, the Natural Resources Defense Council, Union of Concerned Scientists and PSR called on President Reagan and Congress to 'declare an immediate two-year moratorium on the further production of plutonium for nuclear weapons'. They also challenged 'the Soviet Union to negotiate a bilateral, verifiable cut-off of the production of plutonium – as well as highly enriched uranium – for nuclear weapons'.[38] The overall goal of the group was arms control and limitation; however, the concerns this time were with the damage already done to the environment and the health of Americans as a result of weapons production. In PSR President Christine Cassel's Plutonium Challenge statement claimed that the arms race presented Americans with 'unacceptable risks' to their health and called the health threats posed by weapons production the real 'national security' threats, not fear of Soviet aggression.[39]

PSR also went out on its own, sending a letter to Reagan at the end of 1988, which the White House forwarded onto the DOE for an official response. In this letter, PSR declared that the poor health and safety record at DOE reactors constituted a 'national public health emergency'. The group called for the creation of a 'National Review Commission on Nuclear Weapons Production and Public Health' with the mandate 'to assess the medical, public health, occupational and environmental health consequences of the Department of Energy's operation of the entire U.S. nuclear weapons production, testing, and research industry'. This massive undertaking was to be done without interference or input from the DOE. Instead, PSR recommended that appropriate scientific authorities from agencies like the NAS, the NAS/Institute of Medicine, and the US Public Health Service take charge of analysing the health data. PSR argued that there was a 'profound conflict of interest' at the DOE, which was charged with the dual tasks of satisfying national demands for nuclear weapons production and protecting the health and safety of its employees and the public.[40]

The subsequent correspondence between PSR and the DOE reveals how much PSR's activism had changed between 1986 and 1988. Despite the DOE's detailed letter defending the health and safety record of the DOE-run reactors, PSR confidently maintained its position that there needed to be an independent study of worker health and safety data, while also refuting many of the DOE's claims with regard to its past and improving environmental, health and safety record.[41] The doctors were careful not to overstep their role in the process and recommended the Department of Health and Human Services (HHS) as the appropriate body to undertake the analysis of the DOE's worker health and safety records. However, the group was confident enough to refute the DOE's claim that radioactive releases at its nuclear facilities had 'not resulted in detectable health effects in surrounding populations', arguing this was not a position which could be supported 'from a medical point of view'.[42] Through these kinds of engagements with the DOE, as well as its increasing visibility among important members of Congress, PSR established itself as an important player in the debate over how to manage the legacy of environmental contamination and disregard for public health within the American nuclear weapons complex. Indeed, following on the heels of this exchange, Cassel was invited by Argonne National Laboratory (one of the DOE-run laboratories) to become part of the working group 'involving the creation of a Comprehensive Epidemiologic Data Resource (CEDR)', containing information about all DOE worker health and safety data.[43]

By 1991, the DOE finally conceded to public appeals and signed a 'Memorandum of Understanding,' agreeing to transfer research and administrative control of epidemiological studies, including dose reconstructions and exposure assessment studies, to the Department of Health and Human Services (HHS).[44] While PSR was certainly not the only group actively campaigning for the release of this data to HHS, it played an integral role in seeing the issue through to the end and keeping it on the agendas of both HHS and Congress. For instance, Geiger made a statement before the DOE-created Secretarial Panel for the Evaluation of Epidemiological Research Activities (SPEERA), urging the panel to reaffirm PSR's call to release the DOE health and safety data.[45] Several PSR members across the country also participated in local inquiries into the health implications of the nuclear weapons complex. Among them was Thomas Hamilton, who was an MD with both a Masters and a PhD in Public Health. A relatively new PSR recruit in 1989, Hamilton was one of the co-investigators of the prestigious CDC-sponsored Hanford Thyroid Disease Study.[46] PSR also worked tirelessly to keep important members of

Congress and HSS informed about the proceedings of the SPEERA panel and DOE plans to transfer the health data.[47] In an update on PSR's activism in this area, PSR staffer, Todd Perry reported,

> I have been advised by various Congressional and HHS staff professionals that it would be useful for ...[PSR] to send a letter to appropriate DOE and HHS staffs...to let them know who we are and that we are following the [Memorandum of Understanding] and CEDR processes. It is clear that if PSR and a few other concerned individuals were not monitoring these issues, DOE and HHS would have no incentive to do this work in the first place.[48]

It is clear that PSR's continued work in this regard clearly helped keep these issues on the Congressional register and influenced DOE Secretary Watkins' decision to issue the 1991 DOE Memorandum.

11.4   IPPNW joins the environmental opposition

It was not until December 1988 that IPPNW made a commitment to raising awareness about the environmental and public health consequences of the nuclear weapons complex. Several factors explain the small lag between PSR and IPPNW's shift towards environmental issues, but the primary reason was that IPPNW was a Cold War creation, which by 1987 united over 50 national affiliates in a struggle to achieve a Comprehensive Test Ban between the superpowers. While global environmental concerns had received increasing attention by international NGOs and governments around the world since the early 1970s,[49] the Cold War and preoccupations with military aggression were a barrier to efforts at expanding the 'traditional definition of security' to include protection from environmental contamination.[50] Accidents like the Bhopal chemical disaster at the Union Carbide Plant in 1984 and the meltdown at the Chernobyl commercial nuclear power plant in 1986 reminded governments and international activists that risks from chemical and radioactive contamination could not be contained within national borders. But as Michelle Zebich-Knos illustrates, it was not until the Cold War ended that 'advocates of extended security were able to garner increasing attention as they made their case for environmental security concerns'.[51] For this reason, the CTB remained IPPNW's focus until the end of 1988, despite its initial reaction to Chernobyl.

Perhaps IPPNW foretold the Cold War's demise, but it is more likely that the group was experimenting with ways to captivate concerned citizens, who had become numb to the threats of nuclear war. Public malaise intensified with the collapse of the Soviet Union in November of 1989 and people were generally happy to ignore the issue of nuclear disarmament. By shifting one year earlier towards a focus on the environmental and public health legacy of the nuclear arms race, IPPNW reaffirmed its status as a key player in the anti-nuclear movement. Since the group had always been concerned about national and international security, expanding its definition of security to include an analysis of the ways that environmental degradation from the arms race posed an additional and more pressing international security threat was quite seamless. The fact that PSR had joined American environmentalists to raise awareness about the environmental degradation caused by the nuclear weapons complex after Chernobyl put IPPNW in an even better position to redefine its approach to activism.

The international organisation unveiled its plans to join this growing environmental, public health movement in December 1988, when the group formed the International Commission to Investigate the Health and Environmental Effects of Nuclear Weapons Production. Its mission was to 'describe' to the general public in 'scientific' yet 'accessible' terms the health and environmental price of the nuclear arms race, with the goal of inspiring a widespread group of people to further action.[52] The commission concluded with three publications: *Radioactive Heaven and Earth* (1991), which focused on the health and environmental impact of nuclear weapons testing; *Plutonium* (1992), which focused on the many hazards of plutonium production and high level radioactive waste disposal; and *Nuclear Wastelands* (1995), which was a global guide to the health and environmental effects of nuclear weapons production. IPPNW built on PSR's earlier work and applied it to the international arena. Although there were initial tensions between PSR and IPPNW over the latter's decision to shift its focus, which PSR interpreted as a power grab,[53] the two groups worked out their differences and the International Commission continued. Its publications, moreover, assigned credit to PSR for blazing the trail for IPPNW and also for generously sharing resources from its own studies in the US.

Something else happened in 1989 to further embolden IPPNW in its public health and environmental activism. In 1990, the US National Research Council's Fifth Commission on the Biological Effects of Ionising Radiation (BEIR V) made it fashionable to talk about previously controversial topics like the risk of cancer and genetic defects from low-level radiation exposures. BEIR V, which was initially announced in the American news media in late 1989, significantly revised the findings of BEIR III (1980), claiming that the risks of 'getting cancer after being exposed to a low dose of radiation [wa]s three to four times higher' than the 1980 estimate.[54] The overall message of the report was that, while ordinary citizens need not be alarmed, there was no known safe level of radiation exposure.[55] Experts who made this argument in court prior to 1990 were dismissed because it was accepted that extremely low levels of radiation posed little to no health risk. By making this one-time 'extreme view' become 'mainstream,' it had an immediate impact on the work of the IPPNW Commission and shaped its study of nuclear weapons testing.[56] In *Radioactive Heaven and Earth,* the authors boast that their most significant contribution to the study of weapons testing is their use of the 1990 BEIR report to recalculate 'the number of cancer cases and deaths expected from global scattering of fallout'.[57] Similarly, PSR freely discussed the relationship between low and high level radiation exposures and cancer and genetic defects in its work to release the DOE worker health and safety data after 1989. Whereas both groups had shied away from discussing the health effects of low-level radiation exposure since the 1970s, it is fair to say that both groups were comfortably immersed in these issues after 1989.[58] This, in particular, underlines the significance of the shift in the activism of both groups.

While these external and internal factors prepared IPPNW to embark on this next phase of activism, there was still an important element missing. PSR President Cassel identified it in a letter to Lown. In its initial press release, announcing the International Commission, IPPNW declared its intentions to 'STUDY the health effects of nuclear weapons production facilities'. Cassel called this statement both 'unbelievable and irresponsible'. She asked,

> Who are the people with IPPNW who are going to be able to con-
> duct an epidemiologic study of that magnitude, and how will it
> be funded? This seems to me to be an outrageous claim which dis-
> credits the work of all of us. When PSR had its press conference in
> Washington a few weeks ago, we were very cautious to call for the
> appropriate medical organization in the United States (the CDC,
> IOM, AMA, APHA, NIOSH, etc.) to do this work and not to make
> flighty and unsubstantiated claims that we would do it ourselves.[59]

Cassel's criticisms reinforce my argument that relevant biomedical expertise was not IPPNW or PSR's strength. This was the case when the groups were working on the CTB and it remained so when IPPNW announced these new plans.

The group did three things to remedy this shortcoming. First, it formed a partnership with scientists at the Institute for Energy and Environmental Research (IEER). Arjun Makhijani, who was IEER's president, had a PhD in engineering with a specialty in nuclear fusion. The first publication of the commission, *Radioactive Heaven and Earth*, was written almost exclusively by Makhijani[60] and IEER acted as the chief scientific consultant for the entire body of work produced by the commission. Unlike IPPNW, this small non-profit organisation had extensive experience 'assessing the environmental problems arising from nuclear weapons production' and actually performed 'many analyses of plants in the U.S. nuclear weapons complex' prior to its work with IPPNW.[61] This partnership gave IPPNW the critical edge it lacked, enabling the group to make a meaningful contribution to this new phase of anti-nuclear activism. In addition, IPPNW replaced Robbins as director of the commission and hired Dr Howard Hu in 1991. While both Robbins and Hu were physicians with masters degrees in public health, Hu had the additional qualification of a PhD in epidemiology, with a focus on environmental epidemiology. This, then, better equipped IPPNW to embark on an epidemiological assessment of the environmental and public health impact of nuclear weapons production. Likewise, the commission's collaboration with a wide variety of experts, with specialties in epidemiology, biometrics, ecology, occupational and environmental medicine, and nuclear engineering for *Nuclear Wastelands* ensured that IPPNW put out a solid piece of work that achieved its mission to 'describe' to the general public in 'scientific' yet 'accessible' terms the health and environmental price of the nuclear arms race.[62]

11.5 Conclusion

By 1991, both PSR and IPPNW's work raising awareness about the environmental and public health consequences of the nuclear weapons complex had so firmly rooted the two in the environmental movement that they both considered branching out further to raise awareness about non-nuclear environmental and public health issues.[63] In reality, only PSR developed a sustained commitment to a range of environmental/public health issues related to global climate change and environmental pollution.[64] Nevertheless, between 1986 and 1989 both groups made impressive strides in their abilities to effectively address the immediate connections between militarism, the environment and human health, as they related to the nuclear weapons complex. I have shown throughout this article that while the record of disarmament activism for both groups was essential to their contribution, external factors were central in helping them successfully navigate this transition. PSR was emboldened not only by Chernobyl, but also by the growth of grassroots activism in the US that exposed the poor environmental record of DOE-run reactors, a movement that began before the Chernobyl meltdown. IPPNW's status as an international umbrella organisation that formed around a partnership between American and Soviet physicians concerned about the escalating nuclear arms race meant that it was more firmly entrenched in the Cold War political climate. It is for this reason that the group did not join PSR in raising awareness about the links between the arms race, environmental destruction and human health until the Cold War ended and an international environmental movement, which linked radioactive toxicity with concerns about national and international security, emerged in its wake. Likewise both groups were emboldened by the 1990 release of the US National Research Council's Fifth Commission on the Biological Effects of Ionising Radiation (BEIR V), which significantly revised its conclusions about the risk of cancer and genetic defects from low-level radiation exposures. These factors, in addition to the strategic decision both groups made to form relationships with other experts in nuclear technology, radiation health and safety, ecology and so on, significantly aided their transition to environmental/public health activism.

Notes

1. Swarthmore College Peace Collection (SCPC), PSR Fonds, DG 175, Series 1, Box 11, PSR Statement on Chernobyl, 'Notes and Thoughts for a Letter to the Post on Chernobyl', 9 May 1986.
2. SCPC, PSR Fonds, DG 175, Series 1, Box 11, PSR Statement on Chernobyl, 'Statement from Physicians for Social Responsibility on the Soviet Nuclear Power Plant Accident', 5 May 1986.
3. SCPC, PSR Fonds, DG175, Series 1, Box 11, PSR Statement on Chernobyl, 'Post-Mortem of Monday May 5 Press Briefing: Draft', no date.
4. SCPC, PSR Fonds, DG175, Series 1, Box 11, PSR Statement on Chernobyl, 'Press Activities Around the Chernobyl Incident', 7 May 1986. See also, 'Aide offer declined by Soviet leader of anti-nuclear group', New York Times, 1 May 1986; P. M. Boffey, 'Aides say radioactivity has arrived in the US', New York Times, 6 May 1986; T. Wicker, 'A New Attitude?' New York Times, 9 May 1986; and F. Butterfield, 'U.S. foes debate nuclear strategy', New York Times, 14 May 1986.
5. Francis Countway Library of Medicine, Rare Books and Special Collections (FCLMRBSC), IPPNW Fonds, MC408, Box 54, 12 August 1986 – Correspondence from Peter A. Zheutlin, Director of Public Affairs, IPPNW to Dr. William Foege.
6. T. Wicker (1986). 'The invisible shadow', New York Times, 3 June, A27.

7. Other topics discussed at the conference included IPPNW's efforts to establish a health study of the 100,000 evacuees living within a 30 kilometre radius of the Chernobyl meltdown site, along with the growing concerns about the biological effects of weapons testing in the Pacific region. FCLMRBSC, IPPNW Fonds, MC408, Box 54, 'Physicians meet in West Germany', *Fiji Sun*, 7 June 1986.

8. SCPC, PSR Fonds, DG 175, Series 1, Box 25, PSR Materials 1960s and 1970s (1962). 'Physicians for social responsibility', *New England Journal of Medicine*, 266, p. 361.

9. See, for example, 'Study discounts survival plans: Doctors cite inadequacy of facilities in A-war', *New York Times*, 31 May 1962, p. 7; 'Doctors denounce shelter program', *New York Times*, 25 June 1963, p. 11; and L. Grinspoon, MD, 'Fallout shelters and mental health', *Medical Times*, June 1963, found in SCPC, PSR Fonds, DG 175, Series 1, Box 25, PSR Materials 1960s and 1970s.

10. Prepared by Special Study Section, Physicians for Social Responsibility (1962), 'Symposium: The medical consequences of thermonuclear war', *New England Journal of Medicine*, 266 (May 31), pp. 1126–55, 1174.

11. H. Caldicott (1996). *A Desperate Passion: An Autobiography* (New York: W.W. Norton & Company), p. 160.

12. H. Caldicott, *A Desperate Passion*, pp. 198–9.

13. SCPC, PSR Fonds, DG 175, Series 1, Box 25, PSR Annual Reports, '81–83, 'Physicians for social responsibility: Annual report, 1982', p. 2.

14. See, for example, E. Chivean, MD, S. Chivean, R. J. Lifton, MD, J. E. Mack, MD (eds) (1981). *Last Aid: The Medical Dimensions of Nuclear War* (New York: W.H. Freeman and Company).

15. H. J. Geiger, 'The medical effects on a city in the United States', in *Last Aid*, p. 141.

16. See, for example, P. J. Lindop, and J. Rotblat, 'The consequences of radioactive fallout', in *Last Aid*, pp. 249–79.

17. SCPC, PSR Fonds, DG175-Acc. 94A-073 – Series 1, Box 4, File 9, Karen Dorn Steele, 'U.S., soviet doctors building bridges', *Spokane Review*. 4 October 1987.

18. Library and Archives of Canada (LAC), Bertell Fonds, MG31K39, Volume 11, FILE 11 – BOSTON EDISON CO., PART 1, 'United States of America nuclear regulatory commission: In the matter of Boston Edison Company et al.' (Pilgrim Nuclear Power Station Unit 2), Docket No. 50-471. Tuesday, 19 April 1977, 7151–83.

19. 'Statement from physicians for social responsibility on the Soviet nuclear power plant accident', 5 May 1986.

20. L. Nash (2003). 'Finishing nature: Harmonizing bodies and environment in late-nineteenth-century California', *Environmental History*, 8, 1, 25.

21. On industrial hygiene, see, C. Sellers (1999). *Hazards of the Job: From Industrial Disease to Environmental Health Science* (Chapel Hill: University of North Carolina Press) and J. C. Burnham (1995). 'How the discovery of accidental childhood poisoning contributed to the development of environmentalism in the United States', *Environmental History Review*, 19, 3, pp. 57–81. On regulating radiation health and safety, see, B. Hacker (1987). *The Dragon's Tail: Radiation Safety in the Manhattan Project, 1942–1946* (Berkeley, CA: University of California Press); J. E. William (1999). 'Donner laboratory: The birthplace of nuclear medicine', *The Journal of Nuclear Medicine*, 40, 1, 16N, 18N, 20N; J. Heilbron and R. Seidel (1990). *Lawrence and his Laboratory: A History of*

the Lawrence Berkeley Laboratory (Berkeley: University of California Press); D. S. Jones and R. L. Martensen (2003). 'Human radiation experiments and the formation of medical physics at the University of California, San Francisco and Berkeley, 1937–1962', in J. Goodman et al. (eds) *Useful Bodies: Humans in the Service of Medical Science* (Baltimore, MD: Johns Hopkins University Press); C. Caufield (1989). *Multiple Exposures: Chronicles of the Radiation Age* (Toronto: Stoddart), and C. Clarke (1997). *Radium Girls: Women and Industrial Health Reform, 1910–1935* (Chapel Hill: University of North Carolina Press).

22. For a discussion of this trend in the post-war period, see B. Balogh (1991). *Chain Reaction: Expert Debate and Public Participation in American Commercial Nuclear Power, 1945–1975* (Cambridge: Cambridge University Press). Both Christopher Sellers and Linda Nash document the declining significance of physicians to industrial hygiene in the post-World War II years. C. Sellers (1994). 'Factory as environment: Industrial hygiene, professional collaboration and the modern sciences of pollution', *Environmental Health Review*, 18, p. 1, special issue on Technology, Pollution, and the Environment (Spring 1994), p. 73; L. Nash (2004). 'The fruits of ill-health: Pesticides and workers' bodies in post-World War II California', *Osiris*, 2nd series, p. 19, *Landscapes of Exposure: Knowledge and Illness in Modern Environments*, 205.

23. There were three physicists, two physicians and one physician/physicist. S. Boudia (2007). 'Global regulation: Controlling and accepting radioactivity risks', *History and Technology*, 23, 4, pp. 391–2.

24. L. Rumiel (2009). 'Random murder by technology: The role of scientific and biomedical experts in the anti-nuclear movement, 1969–1992' (PhD diss., York University).

25. Email correspondence with the author, 27 July 2009.

26. See J. Samuel Walker, *Permissible Dose*, pp. 80–90.

27. T. R. Fehner and F. G. Gosling (1996). 'Coming in from the cold: Regulating U.S. Department of Energy Nuclear Facilities, 1942–96', *Environmental History* 1, 2, pp. 13–15.

28. M. Stenehjem Gerber (2002). *On the Home Front: The Cold War Legacy of the Hanford Nuclear Site* (Lincoln and London: University of Nebraska Press, 2nd edn), 2–3. For more on the history of DOE run nuclear facilities and the legacy of environmental contamination in the decades leading up to Chernobyl, see, R. Anderson (1989). 'Environmental, safety and health issues at U.S. nuclear weapons production facilities, 1946–1988', *Environmental Review*, 13, 3/4, conference papers, part one (Autumn–Winter), pp. 69–92.

29. M. Stenehjem Gerber, *On the Home Front*, footnote on pp. 327–8; '1949 test linked to radiation in Northwest', *New York Times*, 9 March, p. 35.

30. See, M. Stenehjem Gerber, *On the Home Front*, footnotes for Chapter 8.

31. M. Stenehjem Gerber, *On the Home Front*, pp. 5–6.

32. 'Statement from physicians for social responsibility on the Soviet nuclear power plant accident', 5 May 1986.

33. For a detailed discussion of the rise and decline of the Freeze Movement, see D. Meyer (1990). *A Winter of Discontent: The Nuclear Freeze and American Politics* (New York: Praeger).

34. SCPC, PSR Fonds, DG175 Series 1, Box 4, File 10, Soviet Physicians Tour – 1987 Press, 'Physicians compare Chernobyl aftermath of nuclear explosion', *Hartford Courant*, 6 October 1987.

35. 'Press activities around the Chernobyl incident', 7 May 1986.
36. See also, SCPC, PSR Fonds, DG 175, Series 1, Box 11, October 1988 PSR DOE Press Release, 'Cost of the arms race includes the cost to public health from nuclear arms production', 30 October 1987.
37. C. Sellers, 'Factory as environment', pp. 56–63.
38. SCPC, PSR Fonds, DG 175, Series 1, Box 10, PU Challenge 87, 'The plutonium challenge: An open letter to the president of the United States', 5 November 1987, p. 2.
39. SCPC, PSR Fonds, DG 175, Series 1, Box 11, PSR Statement on Chernobyl, 'Statement of Christine Cassel, M.D. Regarding the plutonium challenge', 5 November 1987.
40. SCPC, PSR Fonds, DG 175, Series 1, Box 11, File 8, Christine Cassel and Victor Sidel of PSR to President Ronald Reagan, 26 October 1988.
41. SCPC, PSR Fonds, DG 175, Series 1, Box 11, File 8, Correspondence from Ernest C. Baynard, III, Assistant Secretary, Environment, Safety, and Health, at the DOE to Christine Cassel, 25 January 1989; and Cassel and Jack Geiger's response to Bayard, 7 February 1989.
42. Correspondence from Christine Cassel and Jack Geiger to Ernest Baynard, III, 7 February 1989.
43. SCPC, PSR Fonds, DG 175, Series 1, Box 11, CEDR/NAS 89, Correspondence from S. Jay Olshansky, PhD at Argonne National Laboratory to Christine Cassel, 18 July 1989.
44. Michele Stenehjen Gerber, *On the Home Front*, 212.
45. SCPC, PSR Fonds, DG175, Series 1, Box 11, File 19, 'Statement of H. Jack Geiger, M.D. Before the Secretarial Panel for the Evaluation of Epidemiological Research Activities', Chicago, Illinois, 26 October 1989.
46. SCPC, PSR Fonds, DG175, Series 1, Box 11, re: Marshall Islands, Thomas Hamilton, MD, PhD, MPH, 'Testimony on health research studies of the Marshall Islanders, presented before the SPEERA, 29 December 1989.
47. SCPC, PSR Fonds, DG175, Series 1, Box 11, Wirth/Wyden '90 Update, Confidential Memo to Ken Rosenbaum, Office of Congressman Wyden from Todd Perry, 'Update on DOE health research program', 7 May 1990.
48. SCPC, PSR Fonds, DG175, Series 1, Box 11, Wirth/Wyden '90 Update, 'Confidential memo', From Todd Perry to Jack Geiger, Wes Wallace, Tony Robbins and Christine Cassel, 8 May 1990.
49. M. Zebich-Knos (1998). 'Global environmental conflict in the post-cold war era: Linkage to an extended security paradigm', *Peace and Conflict Studies*, 5, p. 1, found on the following web link for George Mason University, www.gmu.edu/academic/pcs/zebich.htm, 1–2, accessed 15 July 2009.
50. M. Zebich-Knos, 'Global environmental conflict in the post-cold war era', p. 2.
51. M. Zebich-Knos, 'Global environmental conflict in the post-cold war era', p. 3.
52. A report of the International Physicians for the Prevention of Nuclear War and Institute for Energy and Environmental Research, *Radioactive Heaven and Earth: The Health and Environmental Effects of Nuclear Weapons Testing In, On, and Above the Earth* (New York: The Apex Press, 1991), p. ix; and A report of the International Physicians for the Prevention of Nuclear War and Institute for Energy and Environmental Research, *Plutonium: Deadly Gold of the Nuclear Age* (Cambridge: International Physicians Press, 1992), p. xii.

53. FCLMRBSC, Bernard Lown Fonds, MC899, Box 8, File: Commission – Correspondence, Letter from Christine Cassel to Bernard Lown, 8 December 1988.
54. SCPC, PSR Fonds, DG 175, Series 1, Box 11, BEIR V/Radiation 89-90. E. Marshall, 'Academy panel raises radiation risk estimate', *Science,* 5 January 1990, 247, pp. 22–3.
55. See also, SCPC, PSR Fonds, DG175, Series 1, Box 11, Glenn/Alvarez Report. R. Jeffrey Smith, 'Low-level radiation causes more deaths than assumed, study finds', *The Washington Post,* 20 December 1989, A3; and P. J. Hilts, 'Higher cancer risk found in radiation', *New York Times,* 20 December 1989.
56. E. Marshall, 'Academy panel raises radiation risk estimate', pp. 22–3.
57. *Radioactive Heaven and Earth,* p. x.
58. See, for example, SCPC, PSR Fonds, DG 175, Series 1, Box 16, PSR Fact Sheets, 'A fact sheet on radiation and health', and 'A fact sheet on ionizing radiation'. Date unknown.
59. FCLMRBSC, Bernard Lown Fonds, MC899, Box 8, File: Commission – Correspondence, Letter from Christine Cassel to Bernard Lown, 8 December 1988.
60. *Radioactive Heaven and Earth,* p. xi.
61. A. Makhijani, H. Hu and K. Yih (eds) (1995). *Nuclear Wastelands: A Global Guide to Nuclear Weapons Production and its Health and Environmental Consequences* (Cambridge: MIT Press), p. xvii.
62. A. Makhijani, H. Hu and K. Yih, (eds), *Nuclear Wastelands,* pp. 641–2.
63. See, for example, FCLMRBSC, Bernard Lown Fonds, MC899, Box 8, Commission on Health and Enviro, 'IPPNW Working Group on the Environment: Recommendations to the Executive Committee', drafted by Howard Hu and Katherine Yih, with the participation of the members of the WGE, 26 September 1992, 1–15.
64. For a full accounting of PSR's commitment to environmental health issues, see the group's website, www.psr.org.

12
The Impacts on Human Health and Environment of Global Climate Change: A Review of International Politics

Ingar Palmlund

> We feel that even when all *possible* scientific questions have been answered, the problems of life remain completely untouched. Of course there are then no questions left, and this itself is the answer.
> Ludwig Wittgenstein. *Tractatus Logico-Philosophicus*, proposition 6.52.[1]

Soaring temperatures. Violent tempests and floods. Arctic and Antarctic ice melting. Glaciers vanishing. Forest fires. Severe drought in some regions, severe, unexpected inundations in others. Sub-Saharan Africa is plagued by drought and increasing desertification and the great African lakes are shrinking. In South-East Asia, increasing periods of drought destroy livelihoods in some areas. In others, among them Bangladesh, the sea gradually claims ever more of the land. Due to changing climate, lives are being lost and more lives will be lost. In all this, not only human lives but also human health is under threat on a large scale – although in different ways in different regions on Earth. Natural disasters, starvation, migration and widening, violent conflicts over territories and resources barely reach the headlines in media in rich countries, unless they can be played up as major dramas or involve threats to important business interests. The changing climate is driving dynamic forces that move, perhaps irreversibly, in a direction threatening human civilisation as we know it.[2] (See Figure 12.1 IPCC (2007) Global and continental temperature change.)

The purpose of this paper is to show that although global climate change has potentially devastating consequences for human lives and health, public health concerns emerged late in the discussions of

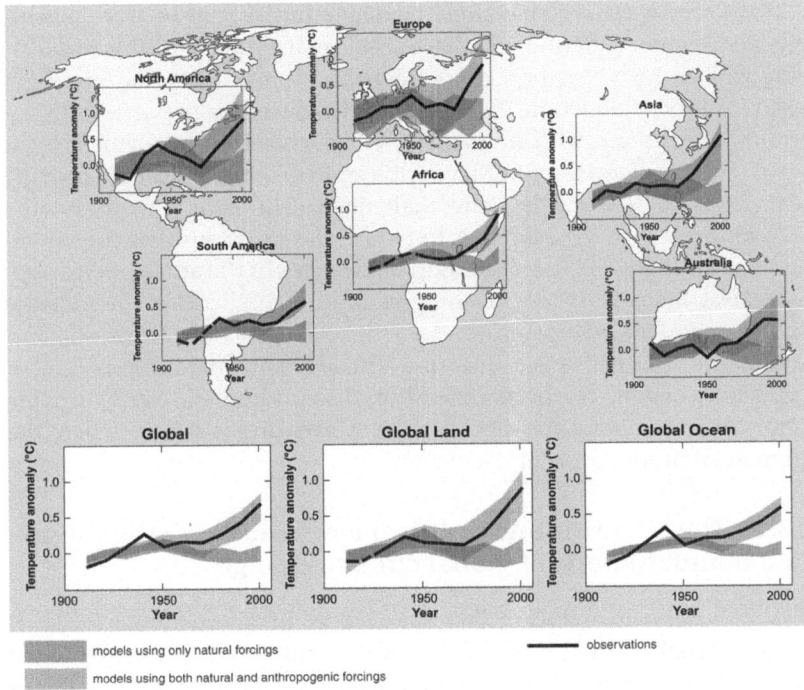

Figure 12.1   IPCC (2007) Global and continental temperature change

international policies and they have not been a strong driving force for action to prevent the deterioration of the human environment. First, the development of the scientific consensus over the causes of climate change and political initiatives to control the hazard is described. How public health concerns have been featured in the international political process over climate change is traced. Finally, the environmental ethics guiding the decision-making on global climate change are highlighted.

Risks to humans and to what humans value are generally regarded as triggers for pre-emptive action and for measures to abate the hazards or at least mitigate their impact. However, in the discourses over global climate change, the threats to human health and, indeed, human survival in many regions have rarely been accorded a high enough profile to inspire drastic action. Shortsighted political manoeuvring over the responsibilities for reducing greenhouse gas emissions have overshadowed the need to protect human health and lives from the impacts of global climate change.

If the scenarios of global climate change turn out to be true, calamities and human suffering on a large scale are inevitable. Rising temperature will, in many regions, cause deserts to spread into land presently used for food production. With over half the world's population living in urban areas and with most major cities located in low-lying coastal areas, the predictions of sea-level rise and ocean acidification provide a scenario of catastrophes. The challenge for humanity, if sustainable development is to be secured, is to reduce the emissions that contribute to the seemingly relentless build-up of the gases in the atmosphere, trapping heat like a glass roof over a greenhouse and causing the increase in average global temperature, droughts, rising sea levels, storms and inundations. Effective action to prevent and mitigate disasters has, so far, hardly begun. Yet, both ancient myths and human history provide many lessons about the necessity of preventing disasters when the human environment is under threat.

12.1   The scientific and political consensus over the causes and manifestations of global climate change

Transboundary air pollution first emerged as an international environmental problem in the late 1960s, when sulphur dioxides, nitrogen oxides and volatile organic compounds – a triad generally known as SO_x, NO_x and VOCs – stemming from the burning of fossil fuels were identified as the causes of acid rain that destroyed flora and micro-fauna in certain regions. The problem was addressed by imposing restrictions on the gas emissions. Technologies developed to prevent, capture or filter the noxious gases made it possible to clean the emissions, so that mainly carbon dioxide (CO_2), believed to be harmless, was emitted. In some countries installing such technology became a mandatory requirement; but in large regions, emissions causing acid rain continued unabated.

The first *global* air pollution hazard identified was the man-made chemical chlorofluorocarbon (CFC), also known as freon, created in 1928 and widely adopted as a non-flammable, supposedly safe chemical in refrigeration, fire suppression systems and manufacturing processes. CFC and related substances were in 1974 identified as gases depleting the stratospheric ozone layer that protects life on Earth against ultraviolet radiation from the sun.[3] In the mid 1980s, the depletion of the stratospheric ozone layer by CFCs – a 'hole' forming above Antarctica[4,5] – was identified as a hazard that would directly affect human health. The specific sources of the implicated gas emissions were easy to locate. Substitutes were identified. International agreements established that the production and use of CFCs should be phased out.

In the late 1980s, concerns over anthropogenic emissions of gases with the potential to cause global climate change emerged as an issue in international politics. A majority in the scientific community had become convinced that many local and regional calamities, such as regional droughts and inundations, could be manifestations of global climate changes caused by emissions of so-called greenhouse gases – primarily CO_2 from the use of fossil fuels, methane (CH_4), nitrous oxide, and CFCs[6] – a global environmental threat emanating from nature as well as from the burning of fossil fuels and industrial activities.

The main anthropogenic greenhouse gas sources are the burning of fossil fuels and deforestation. Land use changes – mainly deforestation in the tropics – account for up to one fifth of total anthropogenic carbon emissions. Other greenhouse gas sources are livestock enteric fermentation and manure management, paddy rice farming, wetland changes and the use of fertilisers. The use of CFCs also contributes. The rising temperatures that accompanied twentieth-century development, based as it was on exploiting the solar energy preserved in coal, oil and gas, are telling a story that is not only about success and progress.

The scientific evidence that human activities contribute to global climate change is robust (see Figure 12.2, IPCC (2007)). The theoretical underpinnings of the science are almost as old as those of Darwin's dangerous ideas about evolution. In 1895 Swedish scientist Svante Arrhenius showed calculations of how changes in the CO_2 content of the atmosphere would affect the surface of Earth, indicating that a doubling of CO_2 emissions would produce a global temperature warming of four to six degrees C.[7] To map air pollution and the build-up of gases in the atmosphere is complex and expensive. Since the mid-twentieth century, Arrhenius' calculations have been substantiated by measurements in air, water and ice, and observations of changes in flora and fauna. The build-up of CO_2 in the atmosphere has since 1958 continually been measured at the Mauna Loa observatory in Hawai, a dramatically rising curve zigzagging between summer and winter measurements, starting at 315 CO_2 parts per million (ppm), by 2009 reaching ca 385 CO_2 ppm, and continuing to rise.[8] Correlation does not mean causation, but this curve, fitted with data on carbon emissions and compared with evidence of temperature changes in ice cores and other sources, demonstrates that, indeed, anthropogenic emissions

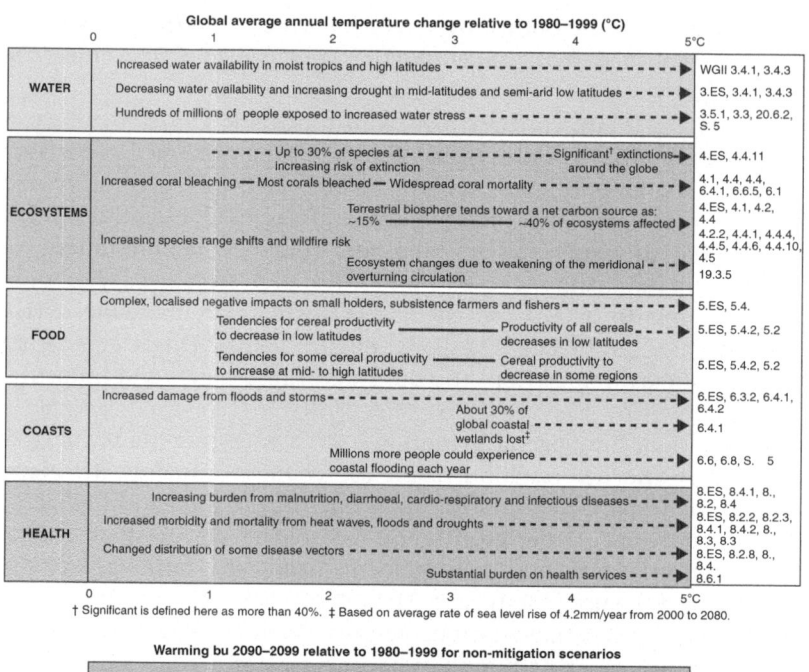

Figure 12.2   IPCC (2007) Examples of impacts associated with global average temperature change. (Impacts will vary by extent of adaptation, rate of temperature change and socio-economic pathway)

of greenhouse gases have driven a rise in average global temperature. A network of measuring stations across the globe, developed in recent decades, has further confirmed that greenhouse gases stemming from human activities are changing the global climate for the foreseeable future.

At the first United Nations (UN) Conference on the Human Environment in 1972, the atmospheric accumulation of CO_2 was announced as one among many other potential future environmental hazards. Five years later, in 1977 at a conference in Villach, Austria held by the International Council for Science's (ICSU) Scientific Committee on Problems in the Environment (SCOPE), climatologists compared data

from their observations with the theory of global warming.[9] Worried
by what they found out, some of these scientists on their return home
warned about what they perceived as a future threat,[10] but it was not
until another climatology conference in Villach in 1985[11] that scientific
opinion seemed sufficiently firm to alert governments to evidence that
global warming appeared to be an inevitable hazard. When these scientists, mainly from the fields of meteorology and atmospheric chemistry
and physics, took the first initiatives to inform political players about
the risks of global climate change, their objective was to eliminate
ignorance. Research would reduce the scientific uncertainty about the
risks.[12] These scientists' world-views and epistemological mindsets came
to define the nature of the issue as a major atmospheric threat with
terrestrial impacts. Human vulnerability was implied but rarely mentioned. Since no immediate consequences for human health and no
evident manifestation of climate change were obvious, political concern
was not aroused.[13]

In 1987, the World Commission of Environment and Development
chaired by Dr Gro Harlem Brundtland, at the time prime minister in
Norway and later director general of the World Health Organisation
(WHO), in its report *Our Common Future* drew attention to major global
issues concerning the human environment that the international policy
community would have to face. The Brundtland Commission pointed
to the environmental impacts of fossil fuel use as one of the crucial
global issues.[14] It also coined a concept, 'sustainable development',
which would set the tone for environmental policy efforts for decades
to come, and it prepared the ground for a second UN Conference on the
environment to be held in Rio de Janeiro, Brazil in 1992. Research in
the US, UK, Sweden, the Netherlands, Japan and other countries gradually strengthened the body of knowledge regarding the influence of the
increasing heat-trapping gas emissions on atmosphere and oceans.[15]
Members of the scientific community, as well as some governments[16]
and non-governmental groups, began to urge that action should be
taken to deal with the threat. Demands were raised that emissions of
greenhouse gases ought to be restricted by an international agreement.
The scientific community was divided. Certain scientists and many
powerful groups in the business community questioned the postulated
relationship between greenhouse gas emissions and climate change,
pointing to inadequate data, claiming that the modelling of future climate development was inaccurate, and that there were no recognised,
generally acknowledged manifestations of climate change. Often supported by the fossil fuels industry, critics have laboured to undermine

the scientific evidence of the harmful effects of fossil fuels.[17] Political pressure has also at times forced a tailoring of presentations to suit political goals.[18]

The industrialised countries have since the early nineteenth century produced much of the greenhouse gases accumulated in the atmosphere. Estimates for 1980 alone, for example, showed that industrialised countries then were responsible for ca 88 per cent of global CO_2 emissions.[19] Nevertheless, scientific as well as political discourses often have focused on the risks of developing countries' future greenhouse gas emissions. A common argument in the scientific and political communities interested in climate change in the late 1980s and early1990 was that the greenhouse gas emissions expected from developing countries were the major threat, as their accelerated economic growth increased their demand for energy.[20]

In the early scientific discourses over global climate change little attention was given to direct threats for human health and lives, a silence that may have helped to delay action. In the UN Assembly, however, small, low-lying island nations began to demand international measures to protect them against global climate change, which for their people and existence posed imminent and present dangers. The UN Environment Program (UNEP) and the World Meteorological Organisation (WMO) in collaboration with ICSU were given a mandate to prepare an international treaty aiming at the control of the causes and effects of climate change,[21] to be ready for final negotiations and signing at the UN Conference on Environment and Development in Rio in 1992. The work was conducted in two groups: an Intergovernmental Panel on Climate Change (IPCC), whose task was to evaluate available scientific evidence, and an International Negotiation Committee (INC), charged with drafting a treaty text. All member countries of the UN and the WMO were invited to appoint participants. The North/South divide in the world soon came to mark the process. Global climate change as an issue raised in international politics had largely been defined and framed by scientists and politicians in the US and Europe, i.e. the industrialised regions, where most of the emissions contributing to climate change had originated. Yet, the early thrust in international policy initiatives was to restrain developing countries from following the development path that had built industrial wealth. This playscript was to be changed. During the preparatory negotiations over a treaty on global climate change the People's Republic of China, India, and certain other countries refused to participate in a treaty unless they were to receive full consideration of their development needs. A widely circulated pamphlet, *Global Warming in an Unequal World*, written by environmental economists at the Center for Science and Environment in New Delhi, India spelled out the argument of the South: the burden of responsibility for greenhouse gas emissions had been unjustly and incorrectly defined by US interest groups; in an equitable approach to global climate change, policies should be based not on greenhouse gas emissions per country but on greenhouse gas emissions per capita.[22] After protracted negotiations, a broad framework convention on climate change was agreed upon. Detailed commitments were to be specified later.

The UN Framework Convention on Climate Change (UNFCCC), acknowledging that change in Earth's climate and its adverse effects were a common concern for humankind and that human activities had substantially contributed to increasing the atmospheric concentrations of greenhouse gases that enhanced the greenhouse effect,[23] was opened for signature at the Rio conference in 1992 and entered into force in 1994. The ultimate objective established for the UNFCCC was to achieve a stabilisation of greenhouse gas concentrations in the atmosphere at a level that would prevent dangerous anthropogenic interference with the climate system. Such a level should be achieved within a time frame sufficient to allow ecosystems to adapt naturally to climate change, to ensure that food production was not threatened and to enable economic development to proceed in a sustainable manner.[24] The UNFCCC explicitly acknowledged that the largest share of historical and current global greenhouse gas emissions had originated in developed countries, that the per capita emissions in developing countries were still relatively low and that the share of global emissions originating in developing countries would have to grow if these countries' social and development needs were to be met.[25]

Nations signing up as Parties to the UNFCCC pledged to protect the climate system for the benefit of present and future generations of humankind, on the basis of equity and in accordance with their common but differentiated responsibilities. The differentiated responsibilities implied that developed countries listed in annexes to the treaty should take the lead in modifying anthropogenic greenhouse gas emissions.[26] The specific needs and circumstances of developing countries should be taken into account. The parties should take precautionary measures to anticipate, prevent or minimise the causes of climate change and to mitigate its adverse effects. Lack of full scientific certainty should not be uses as a reason for postponing cost-effective measures, when there were threats of serious or irreversible damage.[27] Parties to the UNFCCC should launch national strategies for addressing green-house gases and to share information on greenhouse gas emissions, national policies and best practices.[28]

Five years later, in 1997 in Kyoto, Japan the Parties to the UNFCCC agreed on a first protocol to the UNFCCC.[29] This Kyoto Protocol speci-fied developed countries' commitments to the reduction of greenhouse gases and the transfer of technologies to developing countries. Seven years would pass before the Kyoto Protocol had been accepted by a suf-ficient number of nations responsible for a sufficient volume of green-house gas emissions. It set binding targets for 37 industrialised countries and the European community for reducing greenhouse gas emissions, amounting to an average of five per cent against 1990 levels over the five-year period 2008–2012. The reduction targets are likely not to be met by all signatory parties by 2012.[30]

Comprehensive overviews of the state of science and politics con-cerning global climate change were in 2007 presented in reports by the IPCC[31] and the UN Development Program.[32] Expectations for a new legally binding protocol on commitments to prevent climate change were high, when the parties to the UNFCCC met in Copenhagen in 2009 to negotiate commitments after the expiration of the Kyoto Protocol in 2012. Again, the North/South disparity between developed nations, which have been responsible for much of the greenhouse gases in the atmosphere, and developing nations surfaced. The negotiations ended in a tenuous agreement about continuing negotiations, laying down no firm targets for mid- or long-term reductions of greenhouse gases. No deadline was set for establishing a new legally binding pro-tocol from 2012 onwards. Many speakers from developing countries denounced the deal as a sham process fashioned behind closed doors by a club of rich countries and large emerging powers.[33]

The IPCC, working under the oversight of the WMO Executive Council and UNEP Governing Council, has since 1988 continued to assess the growing scientific evidence on how greenhouse gas emis-sions affect the human environment. The IPCC's reports contain reviews and recommendations regarding the state of climate change science, the social and economic impacts of climate change, and the possible response strategies that could be taken into account in the UNFCCC process. It strives to assess the most recent scientific, tech-nical and socio-economic information produced worldwide relevant to the understanding of climate change. The IPCC does not conduct any research, nor does it monitor climate related data or parameters. Since it is an intergovernmental body, its review of science involves both peer reviews by experts and reviews by governments.[34] In 2007 the IPCC produced its fourth assessment. That year it also received the Nobel Peace Price for its work. A fifth assessment is planned for 2014.[35] As part of their commitments to the UNFCCC, many nations have been funding research related to climate developments, thereby contributing to a voluminous body of evidence for the IPCC to assess. Only in 2007, some 4500 climate studies were published, triple the total of ten years earlier.[36]

In its report in 2007 the IPCC affirmed that an average warming of the global climate was unequivocal, evident from observations of increases in global average air and ocean temperatures, widespread melting of snow and ice, and rising global average sea level. Many natural systems were being affected by regional climate changes, particularly by temperature increases. Since pre-industrial times global greenhouse gas emissions due to human activities had grown, with an increase of 70 percent between 1970 and 2004; these emissions were driving global climate change. Most of the observed increase in global average temperature since the mid-twentieth century was very likely due to the increase in greenhouse gas emissions from human activities. Much evidence indicated that in spite of observed climate changes, mitigation policies and related sustainable development practices, the global greenhouse gas emissions would continue to grow over the next few decades. Moreover, continued greenhouse emissions at or above the observed rates would cause further warming and induce changes in the global climate system during the twenty-first century, very likely larger than those observed during the twentieth century.[37] Regional scale changes would include that warming will be greatest over land and most at high northern latitudes. It will be least over the Southern Ocean and parts of the North Atlantic Ocean. There will be a contraction of snow-cover area, increases in thaw depth over most permafrost regions and a decrease in sea ice extent; in some projections, Arctic late-summer sea ice disappears almost entirely by the latter part of the twenty-first century. There will, very likely, be an increase in frequency of hot temperature extremes, heatwaves and heavy precipitation, a likely increase in the intensity of tropical cyclones and perhaps an increase in the number of tropical cyclones globally. There will be a pole-ward shift of extra-tropical storm tracks with consequent changes in wind, precipitation and temperature patterns. Very likely, precipitation will increase in high latitudes and likely decrease in most subtropical land regions. By mid-century, annual river runoff and water availability will increase at high latitudes and in some tropical wet areas; they will decrease in some dry regions in the mid-latitudes and tropics. Many semi-arid areas, e.g. the Mediterranean Basin, the western United States, southern Africa and north-eastern Brazil, will suffer a decrease in water resources due to climate change.[38]

The IPCC pointed to five key vulnerabilities: the risks to unique and threatened ecosystems, including many biodiversity hotspots; the risks of extreme weather events such as droughts, heatwaves and floods; the unequal distribution of impacts and vulnerabilities, with greater vulnerability of specific groups such as the poor and the elderly, in both developing and developed countries; the aggregate impacts of climate change, i.e. higher damages for larger magnitudes of warming; and the risks of so-called large-scale singularities, e.g. that global warming over many centuries will lead to a sea-level rise due to thermal expansion alone, which probably would be much larger than during the twentieth century. Neither adaptation nor mitigation alone would protect against all the impacts of climate change. The IPCC participants believed that much evidence indicated that greenhouse gas emissions could be stabilised at some level by a portfolio of technologies, available or becoming available within decades. Some 60 to 80 per cent of all reductions could come from improved energy supply and consumption efficiencies, and from improved industrial processes.[39]

The prospects that have emerged from the IPCC's assessments of the scientific evidence of climate change are dire for future generations in most regions of the world. Even if the emissions of gases that drive climate change were halted today, the prospects of improving the situation for the generations living at the end of this century or even next century are not good. There is no button to press to stop quickly the global mechanisms set in motion by the greenhouse gases already accumulated and continuing to accumulate in the atmosphere. No single technology or technological substitute exists as yet that would eliminate carbon emissions or capture and store carbon safely and effectively. Although some industry groups and advocates in the public debate have denigrated the scientific evidence on human-induced global climate change, the predicted climate change has also since the late 1980s functioned as an incentive for developing new, energy efficient, 'greener' technologies.

A separation of science and politics on global climate change was institutionalised by setting up the IPCC as a body separate from the political UNFCCC negotiating process. However, the IPCC's assessments of science have not been devoid of political influence, since governments select the participants, and since summaries of the assessments undergo careful redaction before being released to the public. Negotiations in the

UNFCCC process are full-blown exercises in international politics with Earth in the balance. Nations' security – land, productive resources, and position in the world order – is at stake as well as the wellbeing of their inhabitants. For some nations, climate change is challenging their very survival. For others, concerned with dominance and hegemony in the world, the global scope of the issue of global change may have a special appeal.[40] In the process, national security and development priorities have dominated, and fairly little attention has been given to human health concerns.

12.2 Threats to human health and lives

Risks to human health tend to incite political action. It is therefore interesting to contrast the attention to health risks in the two international conventions concerning global gas emissions that alter the sensitive atmosphere and stratosphere of Earth.

The 'ozone hole' in the stratosphere documented in the 1980s, was immediately identified as a health threat – a phenomenon that increased ultraviolet radiation and hence could cause blindness, malignant melanoma and other skin diseases. Politicians moved rapidly to create an international agreement that would stop the production and use of chemicals depleting the stratospheric ozone – the Vienna Convention for the Protection of the Ozone Layer in 1985 and its subsequent Protocols, primarily the 1988 Montreal Protocol on Substances that Deplete the Ozone Layer. A worldwide phasing out of the production and use of CFCs was agreed upon, with developing countries receiving assistance to meet their obligations. The Vienna Convention and its protocols have been successfully implemented, and the 'hole' in the stratospheric ozone layer is now slowly mending.[41]

The Vienna Convention, and even more explicitly its Montreal Protocol, directly addressed the need to protect human health:

> *1985 The Vienna Convention for the Protection of the Ozone Layer:*
> '...*Determined* to protect human health and the environment against adverse effects resulting from modifications of the ozone layer, HAVE AGREED AS FOLLOWS...'.[42]
> *1987 The Montreal Protocol on Substances that Deplete the Ozone Layer:*
> '.... *Being* Parties to the Vienna Convention for the Protection of the Ozone Layer,

Mindful of their obligation under that Convention to take appropriate measures to protect human health and the environment against adverse effects resulting or likely to result from human activities which modify or are likely to modify the ozone layer, ...'[43]

The known health risks of ultraviolet radiation created a sense of urgency that drove the negotiators to agree on nations' common but differentiated responsibilities to halt emissions of CFC and related substances. Action to remedy the situation was taken fairly rapidly. The chemical industry was able to present substitutes for CFC. With relative ease the origins of the harmful emissions could be identified and restrictions imposed.

The situation with regard to global climate change is different. In the negotiations over measures to protect humanity against the effects of global climate change, the scientists and politicians participating in the IPCC and the UNFCCC face a Herculean task. Any effective measures would demand a curtailment of production and lifestyles protected by powerful economic and national interests. No strong advocacy group has engaged in bringing the risks to human health into the negotiations. In the UNFCCC, it was not the protection of human health and life, but nations' economic development and access to energy resources that were seen as the strategically important issues:

1992 The United Nations Framework Convention for Climate Change:
'• ... *Acknowledging* that change in the Earth's climate and its adverse effects are a common concern of humankind...
• *Affirming* that responses to climate change should be coordinated with social and economic development in an integrated manner with a view to avoiding adverse impacts on the latter, taking into full account the legitimate priority needs of developing countries for the achievement of sustained economic growth and the eradication of poverty,
• *Recognizing* that all countries, especially developing countries, need access to resources required to achieve sustainable social and economic development and that, in order for developing countries to progress towards that goal, their energy consumption will need to grow taking into account the possibilities for achieving greater energy efficiency and for controlling greenhouse gas emissions in general, including through the application of new technologies on terms which make such an application economically and socially beneficial,

- *Determined* to protect the climate system for present and future generations, *Have agreed* as follows: ....'[44]

In all present climate change scenarios, not only will national security, society and communities be affected, but families and individuals will come to harm. The impact on human lives will be played out by changing patterns of nutrition, morbidity and mortality. However, the IPCC, mandated to provide the world with assessments of the science concerning climate change and its potential environmental and socio-economic consequences, only gradually included public health implications in their analysis of global climate change impacts. The first IPCC reports in the early 1990s, foundations for developing the UNFCCC, were primarily state-of-the art assessments of available climatologic evidence of temperature changes over time, available data on emissions, and projections based on modelling of what might happen in the future. The potential impacts of climate change were described and response strategies were discussed, but the risks to human health and life were concealed under explanations of climate mechanisms and temperature trends on land and in air, water, and ice.[45,46]

Only in 1993 were the impacts on human health of global climate change publicly addressed in a serious manner, when the medical journal *The Lancet* initiated a series of articles on what medicine in a warming world might entail.[47] Taking into account that alterations in temperature, rainfall and a pattern of extreme weather events would have an impact on the redistribution of old diseases and the emergence of new ones, the authors highlighted the health implications of the scenarios presented in the IPCC's reports. Rising temperatures would increase deaths from cardiovascular and cerebrovascular disease among the elderly. Among the indirect effects would be changes in vector-born diseases such as malaria, lymphatic filariasis, African trypanosomiasis, dengue fever, yellow fever, Chagas' disease and schistomiasis. Climate change affecting coastal areas might increase incidences of cholera and shellfish poisoning. Hunger and malnutrition would be potential consequences of changes in agriculture and land use. The authors of the articles in *The Lancet* called on the WHO to take a key role in coordinating a 'Global Health Watch'.[48]

Two years later the IPCC published its second assessment of climate-change science – a renewed, deepened scientific-technical analysis of the driving forces and impacts of climate change and options for adaptation and mitigation.[49] There was still fairly little input from public health experts. Based on the relatively few studies of health effects that could be related to climate change, the two main conclusions regarding morbidity and mortality were that an increase or severity of heatwaves would cause a short-term increase in illness and death; and climate-induced changes in the geographical distribution of vector-borne infectious diseases and infective parasites would increase or otherwise alter the transmission of infectious diseases.[50]

To analyse the economic and social dimensions of climate change the panel of natural scientists in the IPCC had felt a need to involve also social scientists in their work.[51] Perhaps confident that in this area, too, numerical information would provide an accurate representation of reality, the IPCC selected economists for a working group on economic and social issues. Perhaps the IPCC leadership was ignorant of the reluctance among many social scientists to put faith exclusively in numerical data about social factors; facts and values rarely being entirely separate in social research on environmental hazards.[52] Since the early 1980s, ignited by the controversies over the risks of nuclear power, a vibrant debate had been conducted in the multi-disciplinary research field of risk analysis[53] with contests over methods such as probabilistic risk assessment, 'willingness to pay' and 'statistical life' used in assessments of environmental hazards.[54]

The IPCC's economist working group set out to estimate in monetary terms the impacts of climate change on human health, and the social costs and benefits of investing in carbon control, pointing out that non-market impacts such as risks to human health, risks of mortality and damage to ecosystems formed an important component of the social costs of climate change. They argued that monetary evaluation of such impacts was essential for sound decision-making, because non-market estimates of damage would produce major uncertainties in the assessment of the implications of global climate change for human welfare. Thence, they conducted an estimate of such non-market costs, based on people's willingness to pay for risk reduction, claiming that with this method the risks of climate change would be treated like any other health risk. Their monetary valuation of human life differed from country to country, a statistical life being assessed at a much higher value in rich countries than in poor countries. The IPCC published their results.[55] Needless to say, this proved quite controversial, again widening the North/South divide in international climate-change politics.[56] Critics objected that the IPCC economists' cost-benefit approach was flawed, since it dealt with victims in one country damaged by activities in another country. And, why should the deaths inflicted by the big greenhouse gas emitters – principally the industrialised countries – be valued differently in the perpetrators' and the victims' countries?[57] This detour meant that some years were lost before the IPCC could present a panorama of the potential effects of climate change in terms of morbidity and mortality in human populations.

In the third IPCC report, published in 2001, the summary for policy makers started with a statement emphasising that the knowledge about the social impacts of climate-change was full of uncertainties:

> Natural, technical, and social sciences can provide essential information and evidence needed for decisions on what constitutes 'dangerous anthropogenic interference with the climate system.' At the same time, such decisions are value judgments determined through socio-political processes, taking into account considerations such as development, equity, and sustainability, as well as uncertainties and risk. The basis for determining what constitutes 'dangerous anthropogenic interference' will vary among regions— depending both on the local nature and consequences of climate change impacts, and on the adaptive capacity available to cope with climate change—and depends upon mitigative capacity, since the magnitude and the rate of change are both important.[58]

The section in the IPCC's 2001 report, which presented the impacts of global climate change, adaption to climate change and vulnerability, included a chapter on human health.[59] Little published evidence had by then indicated changes in people's health status that could be related to climate change. Overall, the negative health impacts of climate change were anticipated to outweigh the positive health impacts. Based on a mix of epidemiological studies and predictive modelling, it seemed likely that an increase in the frequency or intensity of heatwaves would increase the risk of mortality and morbidity, principally in older age groups and among the urban poor. Any regional increases in storms, floods, cyclones and other extreme events associated with climate change would cause physical damage and displacement of people. Such events would have adverse effects on food production, fresh water availability and quality, and they would increase the risks of epidemics of infectious diseases, particularly in developing countries. Climate change would also cause deterioration in air quality in many large urban areas; increases in ground-level ozone and other air pollutants that would affect morbidity and mortality. Vector-borne diseases such as malaria, dengue and leishmaniasis, which are sensitive to climate conditions, would extend to higher altitudes and higher latitudes. Changes and variability in climate would affect also other vector-borne infections, i.a. mosquito-borne encephalitis, Lyme disease and tick-borne encephalitis. Changes in surface water quantity and quality would affect the incidence of diarrhoeal diseases. In coastal areas, the warming of seawater might facilitate the transmission of cholera. The envisaged adverse health impacts of climate change would fall disproportionately on the poor in tropical and subtropical countries. Generally, public health infrastructures needed to be maintained and strengthened as part of the measures to adapt to climate change.[60]

Since the health impacts of climate change are potentially huge and many of the most important global killer diseases are highly sensitive to climatic conditions, the WHO has an important role to play in framing the problems and setting an agenda for action. In the 1990s, the WHO's input in international global climate change politics had a low profile, but in 2000 the WHO started to work with the WMO and UNEP to raise awareness of the public health implications of climate change in highly vulnerable regions. The objective was to inform and to support national public health actors to engage in concrete actions.[61] The WHO in 2003 and 2004 published regional and global comparative risk assessments to quantify premature morbidity and mortality due to a range of risk factors, including climate change[62,63,64]: the projected relative risks attributed to climate change would vary by health outcome and region; increases in diarrheal diseases and malnutrition, primarily in low-income populations already experiencing a large burden of disease were expected; so was an increase in cardiovascular disease mortality attributable to climate extremes, especially in tropical regions.[65]

For the IPCC's 2007 report the WHO was well represented in the working group reviewing climate-related effects on human health. This working-group concluded that climate change was already contributing to the global burden of disease and premature death; it had altered the distribution of some infectious disease vectors, altered the seasonal distribution of some allergenic pollen species, and increased heatwave-related deaths. Projected trends indicated an increase in malnutrition and consequent disorders, including those relating to child growth and development. The incidence of death, disease and injury from heatwaves, floods, storms, fires and droughts would increase. The range of some infectious disease vectors would continue to change. Malaria would in some regions contract but expand elsewhere and the transmission season might be changed. The number of people at risk of dengue fever might increase. The burden of diarrheal diseases would increase as would cardio-respiratory morbidity and mortality associated with ground-level ozone. Climate change might bring some benefits to health, including fewer deaths from cold; however, the benefits would probably be outweighed by the negative effects of rising temperatures worldwide, especially in developing countries.

The authors noted that although, in many respects, population health had improved remarkably over the last 50 years, substantial inequalities in health persisted within and between countries. Non-communicable diseases, such as heart disease, diabetes, stroke and cancer accounted for nearly half of the global burden of disease, but communicable diseases still remained a serious threat to public health in many parts of the world despite immunisation programs and other measures.[66]

The World Health Assembly in 2008 placed climate change on its agenda and demanded a more active engagement from the WHO.[67] In the 2009 run-up for the UNFCCC negotiations *The Lancet,* again, drew attention to the magnitude and severity of the challenge for public health. In a report by *The Lancet* and the University College London Institute for Global Health Commission, the changes in the health panorama expected from global climate change were presented. The authors called for a new advocacy and public health movement working for measures to mitigate and adapt to the effects of climate change on human health, emphasising that the indirect effects of climate change on water, food security and extreme climate events were likely to have more serious effects on global health than the direct effects due the increase in vector-borne diseases and heat waves. Climate change would exacerbate global inequity, since it would have greatest impact on the lives of those who have least access to resources and who have contributed least to the causes of climate change. It would challenge the ability of health-care systems to respond effectively, especially in many low-income and middle-income countries with public health services already insufficient to provide effective health care.[68]

The WHO, however, seemed confident. On the WHO website under the rubric 'Global Climate Change and Health', the following statement could be found in July 2009:

> Fortunately, much of the health risk is avoidable through existing health programmes and interventions. Concerted action to strengthen the key features of health systems, and to promote healthy development choices, can enhance public health now as well as reduce vulnerability to future climate change.[69]

It is difficult to trust in this façade of institutional optimism. Too many pictures of reality have informed about the fallibilities of health systems, when it comes to providing adequate public health measures after natural and man-made disasters. It may be sufficient to recall the lot of the victims of the 2005 Hurricane Katrina striking New Orleans in one of the richest countries or the world. Or, to take examples from developing countries in summer 2009: deaths and suffering due to torrential rains, flashfloods and landslides in Bangladesh; a violent typhoon destroying communities on Taiwan and the coast of mainland China with a million people on flight to save their lives; and the silent wrecking of human lives and societies by the slow, inexorable process of desertification in the sub-Saharan region. Such devastating effects of changing climate had so little news value in the richer countries of the world that they hardly reached media headlines. The attention and support for victims of these catastrophes in poor countries can be compared with the public panic epidemics at about the same time in richer countries over threats from uncommon influenza varieties.

For over half a century, in developed countries, infectious diseases have come to be regarded with relative complacency because a battery of medicinal drugs are available to prevent disease or provide cure. However, the balance in the continually evolving symbiosis between human and micro-organism populations is upset in a world, where over half the human population migrates into densely populated urban areas with inadequate fresh water and sanitation services, where people speedily move from continent to continent as never before, and where micro-organisms quickly adapt to new ecological challenges by developing resistance to the remedies invented to protect human health.[70] No change in Earth's ecosystems, locally, regionally or globally, will alter the fact that humans, for better or for worse, are part of the web of life.

12.3   'What can I know? What shall I do? What may I hope?'

Immanuel Kant posed these three questions in his *Critique of Pure Reason*,[71] fundamental not only in philosophy but in facing the ongoing global environmental deterioration that threatens human life and civilisations on Earth. Whereas the knowledge about global climate change, its causes and its progression into the future gradually has accumulated, major disagreements remain about action that needs to be taken to prevent disasters on a large scale. Disagreements also fester about the prospects for the future. Widening gulfs persist between those who profit in the short term from continuing 'business as usual' and those who may become the victims of natural disasters without adequate resources to protect themselves and their families. Ignorance may be part of the problem. The popular mantra 'Save the Planet' is grossly misleading. The planet will remain, regardless of how climates change. The real challenge is to save the human habitat and humanity.

International environmental policies rest on a set of principles that address Kant's questions. These principles were summarised in the Rio Declaration on Environment and Development signed in 1992 by the heads of states assembled at the UN Conference on Environment and Development.[72] The first of these requires that protecting human lives and health should be a priority in international environmental agreements: 'Human beings are at the centre of concerns for sustainable development. They are entitled to a healthy and productive life in harmony with nature'.[73] Many of the Rio Declaration principles are inscribed in international treaties such as the UNFCCC. In the preamble to the UNFCCC the importance of safeguarding intergenerational equity is emphasised. This means that the right to development must be fulfilled so as to meet developmental and environmental needs of present and future generations in an equitable manner.[74] The UNFCCC also emphasises nations' common but differentiated responsibilities to protect the human environment, acknowledging that developed countries have contributed to shaping some of the global environmental deterioration that the world community is facing. They bear a special responsibility in the international pursuit to achieve sustainable development, considering the pressures their societies place on the global environment and the technologies and financial resources they command.[75]

The principle known as the precautionary principle, a 'better safe than sorry' principle, guided governments signing the UNFCCC in 1992. The precautionary approach, to be applied by nations according to their capabilities, means that where there are threats of serious or irreversible damage, lack of full scientific certainty shall not be used as a reason for postponing cost-effective measures to prevent environmental degradation.[76] The impacts of human-induced global climate change are such that it will be too late to act when everything about human-induced global climate change is documented, verified, accepted and incorporated in general knowledge; too late for precautions and even too late to effectively mitigate the predicted disasters.

The IPCC has been charged with reducing scientific uncertainty about the nature of global climate change and with establishing trustworthy, institutionalised knowledge about the nature of the threat, its causes, its impacts and its course into the future. The priority on finding knowledge to reduce ignorance and get a grip on uncertainty rested on a rational assumption: once a problem is defined correctly, it can be solved, and science will make technologies available for solutions. However, the understanding of the mechanisms in global climate change is gradually clarifying that the process may be irreversible and that there are no handy solutions.

An epistemic community[77] has formed around the threat of global climate change. Epistemic communities share an historical *a priori* grounding of knowledge and its discourses in a scientific field and a language. The global climate change epistemic community is multi-disciplinary but with a hard core of natural science-trained participants. Scientists, who study global weather systems and the atmosphere surrounding Earth, have a tradition of transnational collaboration, perhaps unique in the world of science. Since over a century they have elaborated unifying theories and an ease of communication based on scientific and mathematical language. When their concerns over global climate change reached political decision makers, climate science soon developed into a generously funded research field, where careers could be developed. Scientists investigating the mechanisms and impacts of global climate change have shared data and developed models and scenarios, proceeding in the tradition of what for them was normal science.[78]

No equally strong epistemic community has formed around strategies to deal with the public health consequences of global climate change. The epistemic tradition in medicine emphasises attention to the individual patient more than to the large-scale patterns of disease, ill health and threats to humans. For political action, a strong popular consensus on the need for urgent measures is essential. The medical community has not, at least not yet at the time of this writing in 2009, become a vociferous constituency demanding effective political action to develop public health policies for coping with the effects of global climate change.

One obstacle to action may be the specific language and terminology that has developed in the global climate change epistemic community. To cite Wittgenstein: 'What we cannot speak about we must pass over in silence'.[79] In global climate-speak,[80] terms like 'scenario', 'impact', 'vulnerability', 'irreversible', 'mitigation', 'adaptation' and 'adaptive capacity' have become common currency, verbal coins sometimes so worn that they almost lose meaning, even though they carry sinister messages about the future of humanity. Public health vocabulary is rare in global-climate-speak. The sensory and emotional experiences of the prospects of global climate change get suppressed. Non-scientists are impeded from entering the debate. A seasoned science journalist has pointed to the need to liberate the climate change issue from the suffocating embrace of science and to explain it in legal, financial, poetic and ethical languages for people outside the community of global climate change scientists.[81]

In order to cope effectively with the health and livelihood issues embedded in the proceeding global climate change, what can we know, what shall we do, what may we hope? The known prospects for humanity of global climate change should be a call for action. The oath of allegiance for medical professionals in the World Medical Association's Declaration of Geneva means that they pledge to consecrate their lives to the service of humanity, and that the health of their patients should be their first consideration.[82] This places a special responsibility on them to engage in creating strategies to counter the impacts of climate change. When there is a will, there is a way – but is there a strong enough will to deal with the calamities for humankind, so graphically emerging in the climate-change scenarios? The core question in any humanitarian ethos remains: 'What can I do in order to reduce suffering and sustain hope?' The disasters a changing global climate will cause are a challenge not only for politicians, but also for the broader community working to protect human health and life on Earth.

Notes

1. L. Wittgenstein (1981). *Tractatus Logico-Philosophicus*, translated by D. F. Pears and B. F. McGuinness (London: Routledge & Kegan Paul), proposition 6.52, p. 73.
2. UN Environment Program (2009). *UNEP Yearbook 2009: New Science and Developments in Our Changing Environment* (Nairobi: UNEP), pp. 21–30.
3. M. J. Molina and F. S. Rowland (1974). 'Stratospheric sink for chlorofluoromethanes: Chlorine atom-catalysed destruction of ozone', *Nature*, 249, pp. 810–2.
4. J. C. Farman, B. G. Gardiner and J. D. Shanklin (1985). 'Large losses of total ozone in Antarctica reveal seasonal ClOx/NOx interaction. Letter', *Nature*, 315, pp. 207–10.
5. S. Solomon, R. R. Garcia, F. S. Rowland and D. J. Wuebbles (1986). 'On the depletion of Antarctic ozone', *Nature*, 321, pp. 755–8.
6. Intergovernmental Panel on Climate Change (2007). *Climate Change 2007: The Physical Science Basis: Contribution of Working Group 1 to the Fourth Assessment Report of the Intergovernmental Panel on Climate Change* (New York: Cambridge University Press). The Intergovernmental Panel on Climate Change is hereafter referred to as IPCC.
7. S. Arrhenius (1895). 'On the influence of carbonic acid in the air upon the temperature of the ground', Paper presented at the Stockholm Physical Society, 1895.
8. For the Mauna Loa CO_2 measurements, see the so-called Keeling curve at http://www.mlo.noaa.gov/programs/esrl/co2/img/img_mlo_co2_record_2007.gif, accessed 29 July 2009.

9. B. Bolin (1993). Personal communication, interview, January 1993. The conference is documented in *SCOPE Report* No. 13, 1977.
10. B. Bolin (1993). Personal communication, interview, January 1993.
11. W. J. Maunder (1992). *Dictionary of Global Climate Change* (London: UCL Press); 'Villach conference 1985', p. 218.
12. M. Faber, R. Manstetten and J. L. R. Proops (1992). 'Humankind and environment: An anatomy of surprise and ignorance', *Environmental Values*, 1, pp. 217–42.
13. R. M. White (1990). 'The great climate debate', *Scientific American*, 263, July, p. 38.
14. World Commission of Environment and Development (1987). *Our Common Future* (Oxford: Oxford University Press). Submitted to the UN General Assembly as an Annex to document A/42/427 – Development and International Co-operation: Environment.
15. P. D. Jones and T. M. L. Wigley (1990). 'Gobal warming trends', *Scientific American*, 263, August, pp. 84–91.
16. H. Kohl, F. Hophouet-Boigny, G. Evans, M. H. Mubarak et al. (1989). *The Hague Declaration 11 March 1989.*
17. S. Goldenberg (2009). 'Barack Obama's key climate bill hit by $45m PR campaign', *The Guardian*, 12 May. http://www.guardian.co.uk/environment/2009/may/12/us-climate-bill-oil-gas, accessed 12 May 2009.
18. Union of Concerned Scientists (2007). '*Atmosphere of Pressure*', 30 January 2007. http://www.ucsusa.org/scientific_integrity/abuses_of _science/atmosphere-of-pressure.html, accessed 1 August 2009.
19. R. M. White (1990). p. 41.
20. R. M. White (1990). pp. 36–43.
21. UN General Assembly (1988–1991). *Resolutions A/RES/43/53 6 December 1988, A/RES/44/207 22 September 1989, A/RES/45/212 21 December 1990, and A/RES/46/169 19 December 1991 on protection of global climate for future generations.*
22. A. Agarwal and S. Narain (1991). *Global Warming in an Unequal World* (New Delhi: Centre for Science and Environment).
23. *The United Nations Framework Convention on Climate Change* (1992). Preamble. http://unfccc.int/resource/docs/convkp/conveng.pdf, accessed 29 July 2009. The Convention is hereafter referred to as UNFCCC.
24. UNFCCC, Article 2, Objective.
25. UNFCCC, Preamble.
26. UNFCCC, Article 4, Commitments.
27. UNFCCC, Article 3, Principles.
28. UNFCCC, Article 4, Commitments.
29. UNFCCC (1997). *The Kyoto Protocol*, adopted at the third session of the Conference of the Parties (COP 3) in Kyoto, Japan, on 11 December 1997. http://unfccc.int/kyoto_protocol/items/2830.php, accessed 29 July 2009.
30. The Signatory Parties' targets stated in the Kyoto Protocol for greenhouse gas emission by 2012 in percentage of 1990 emissions are as follows: The European Union (EU) 15 members –8%; Canada, Hungary, Japan and Poland –6%; Croatia –5%; the US, which signed but did not ratify the Kyoto Protocol, –7%; New Zealand, Russian Federation, and Ukraine 0%; Norway +1%; Australia +8%; Iceland +10%.

31. IPCC (2007). *Climate Change 2007: Synthesis Report. Summary for Policymakers*. (Geneva: World Meteorological Office).
32. UN Development Program (2007). *Human Development Report 2007/2008: Fighting Climate Change: Human Solidarity In A Divided World* (London: Palgrave Macmillan).
33. A. C. Revkin (2009). 'A grudging accord in climate talks' *The New York Times*, December 20, p. A1.
34. IPCC (2006). 'Principles governing IPCC work', Approved 1 October 1998, amended 3 and 6–7 November 2003, and 26–8 April 2006. http://www.ipcc. ch/organization/organization.htm, accessed 30 July 2009.
35. IPCC (2009). Fifth Assessment Report (AR5). http://www.ipcc.ch/activities/ activities.htm#1, accessed 21 August 2009.
36. A. C. Revkin (2009). 'Nobel halo fades fast for climate change panel', *The New York Times*, 4 August.
37. IPCC (2007). *Climate Change 2007: Synthesis Report. Summary for policymakers, Contribution of Working Groups I, II and III to the Fourth Assessment Report of the Intergovernmental Panel on Climate Change [Core Writing Team, Pachauri, R.K. and Reisinger, A. (eds)]. IPCC, Geneva, Switzerland*, pp. 2, 5, 7.
38. IPCC (2007). p. 8.
39. IPCC (2007). pp. 18–20.
40. J. M. Broder (2009). 'Climate change seen as threat to US security', *The New York Times*, 9 August, p. 1.
41. UNEP (2009). *Vital Ozone Graphics*: Resource kit for journalists. http://www. grida.no/publications/vg/ozone/page/1379.aspx to /page1396.aspx, accessed 28 July 2009.
42. UN Environment Program (1985). *The Vienna Convention for the Protection of the Ozone Layer*, Preamble.
43. UN Environment Program (1987). *The Montreal Protocol on Substances that Deplete the Ozone Layer*, Preamble.
44. UNFCCC, Preamble.
45. IPCC (1990). *Climate Change: The IPCC Scientific Assessment* (Cambridge: Cambridge University Press).
46. IPCC (1992). *Climate Change: Supplementary Report to the IPCC Scientific Assessment* (Cambridge: Cambridge University Press).
47. D. Sharp and P. R. Epstein (eds) (1994). *Health and Climate Change* (London: The Lancet Ltd.). (Original articles in the *Lancet* in 23 October–11 December 1993).
48. A. Haines, P. R. Epstein and A. J. McMichael, on behalf of an international panel (1994). 'Global health watch: Monitoring aspects of environmental change', in D. Sharp and P. R. Epstein (eds) *Health and Climate Change* (London: The Lancet Ltd.), pp. 31–4.
49. IPCC (1996). *Second Assessment Report: Climate Change 1995 (SAR)* (Cambridge: Cambridge University Press).
50. A. J. McMichael, M. Ando, R. Carcavallo, P. Epstein, A. Haines, G. Jendritzky, L. Kalstein, R. Odongo, J. Patz and W. Piver (1996). 'Human population health', in IPCC (1996). *Climate Change 1995: Impacts, Adaptations, and Mitigation of Climate Change: Scientific-Technical Analyses. Contribution of Working Group II to the Second Assessment Report of the Intergovernmental Panel on Climate Change* (Cambridge: Cambridge University Press), pp. 561–84.

51. B. Bolin (1993). Personal communication, interview, January.
52. N. A. Ashford (1988). 'Science and values in the regulatory process', *Statistical Science*, 3, 3, pp. 377–83.
53. *Risk Analysis: An International Journal*, published on behalf of the Society for Risk Analysis since 1981 represents the field.
54. S. Krimsky and D. Golding (eds) (1992). *Social Theories of Risk*, chapters by i.a. P. Slovic, H. Otway, W. R. Freudenberg, S. O. Funtowicz and J. R. Ravetz, and B. Wynne (Westport, Co and London: Praeger).
55. IPCC (1996). *Climate Change 1995: Economic and Social Dimensions of Climate Change*: *Contribution of Working Group III to the Second Assessment Report of the Intergovernmental Panel on Climate Change* (Cambridge: Cambridge University Press), pp. 9–11, 145–224.
56. US Environmental Protection Agency, National Center for Environmental Economics (1996). 'Correction on Global Warming Cost Benefit Conflict', *EDV-CBN Newsletter*, III, 1.
57. M. Grubb (2005). 'Stick to the target', *Prospect Magazine*, 114, 25 September.
58. IPCC (2001). *Climate Change 2001: Synthesis Report*, Summary for policy-makers, p. 2.
59. A. J. McMichael and A. Githeko et al. (2001). 'Human health', in *Climate Change 2001: Impacts, Adaptation and Vulnerability. Report of Working Group II of the IPCC* (Cambridge: Cambridge University Press), pp. 451–85.
60. IPCC (2001). *Summary for Policymakers: Climate Change 2001: Impacts, Adaptation and Vulnerability. Report of Working Group II of the IPCC* (Cambridge: Cambridge University Press), p. 12.
61. WHO (2006). *Climate Change and Health* (Geneva: World Health Organisation) http://www.who.int/globalchange/climate/en, accessed 20 August 2009.
62. D. H. Campbell-Lendrum, C. F. Corvalan and A. Prüss-Ustün. (2003). 'How much disease could climate change cause?' In: A. J. McMichael, D. H. Campbell-Lendrum, C. F. Corvalan, K. L. Ebi, A. Githeko, K. Ebi, J. D. Scheraga and A. Woodward (eds.). *Climate Change and Human Health: Risks and Responses* (Geneva: WHO/WMO/UNEP), pp. 133–59.
63. M. Ezzati, A. Lopez, A. Rodgers and C. Murray (eds) (2004). *Comparative Quantification of Health Risks: Global and Regional Burden of Disease due to Selected Major Risk Factors, Vols. 1 and 2* (Geneva: World Health Organisation).
64. A. J. McMichael, M. McKee, V. Shkolnikov and T. Valkonen (2004). 'Mortality trends and setbacks: Global convergence or divergence?' *Lancet*, 363, pp. 1155–9.
65. U. Confalonieri, B. Menne, R. Akhtar, K. L. Ebi, M. Hauengue, R. S. Kovats, B. Revich and A. Woodward (2007). 'Human health', in M. L. Parry, O. F. Canziani, J. P. Palutikof, P. J. van den Linden, C. E. Hanson (eds) (2007). *Climate Change 2007: Impacts, Adaptations and Vulnerability. Contribution of Working Group II to the Fourth Assessment Report of the IPCC* (Cambridge: Cambridge University Press), p. 407.
66. Confalonieri et al. (2007). pp. 393–94.
67. A. Costello et al. (2009). 'Managing the health effects of climate change', *Lancet*, Vol. 373, 16 May, p. 1723.
68. A. Costello et al. (2009). 'Managing the health effects of climate change', *Lancet*, Vol. 373, 16 May, pp. 1693–733.

69. http://www.who.int/globalchange/en/, accessed 29 July 2009.
70. W. H. McNeill (1998). *Plagues and Peoples* (New York: Knopf Doubleday).
71. I. Kant (1781/1787). *Kritik den reinen Vernunft (Critique of Pure Reason)* as cited in M. Faber, R. Manstetten and J. L. R. Proops (1992). 'Humankind and environment', *Environmental Values* 1, pp. 217–42.
72. The UN Conference on Environment and Development, *Rio Declaration on Environment and Development*, Rio de Janeiro 3–14 June 1992.
73. *Rio Declaration*, Principle 1.
74. *Rio Declaration*, Principle 3.
75. *Rio Declaration*, Principles 6 and 7.
76. *Rio Declaration*, Principle 15.
77. M. Foucault (1970). *The Order of Things: An Archeology of the Human Sciences* (New York: Pantheon books).
78. T. Kuhn (1970). *The Structure of Scientific Revolutions*, 2nd edn (Chicago: The University of Chicago Press), pp. 35–42.
79. Wittgenstein (1981). *Tractatus Logico-Philosophicus*, Proposition 7, p. 74.
80. W. J. Maunder (1992). *Dictionary of Global Climate Change*, cited above is an example of how an epistemic community's shared language may be confirmed by issuing a special lexicon.
81. F. Pearce (2009). 'No credit where it's due', *Guardian Review*, 9 May. http://www.guardian.co.uk/books/2009/may/09/scienceandnature-climate-change, accessed 9 May, 2009.
82. The Second General Assembly of the World Medical Association (1948). *Declaration of Geneva*, revised May 2006.

13

Epilogue: The Co-Benefits for Health of Meeting Global Environmental Challenges

Paul Wilkinson

This book has examined the changing nature of environmental public health in the last 150 years. Until the last quarter of the twentieth century, the agenda had been dominated by concerns about chemical contamination of local environments, about health problems of urbanisation and of persisting threats of infectious disease. In many instances these were the products of industrial and agricultural development, shaped by social and political contexts.

What has emerged in recent years is the recognition of new large-scale environmental concerns with capacity to have a profound effect on the Earth's natural systems at a global level – through depletion of natural resources, reduced biodiversity and accelerated species loss, and alteration of climatic conditions.

Global climate change is the most recognised but by no means the only example of major global-scale environmental threats, which have followed in the wake of stratospheric ozone depletion. Such problems are different in a number of respects from more traditional environment and health concerns. Their effect is not the product of local toxic contaminants, but of disturbances of systems from the accumulative impact of seemingly relatively benign emissions or of the depletion of natural resources, insignificant at individual level, that become problematic when multiplied up to regional or global scales. They also differ from traditional local environmental concerns in their visibility to the public.

Climate change is still largely a problem predicted for the future rather than an issue with obvious impact today, and its likely health, social and environmental consequences remain uncertain in many details (McMichael, Woodruff et al., 2006). Our concern about it comes second-hand from the models and reports of panels of scientific experts rather than from our own direct experience.

270

Addressing climate change will require actions on two levels: actions to reduce the vulnerability of natural and human systems against that component of climate change which is already inevitable; and efforts to reduce the intensity of global warming by reduced emissions of greenhouse gases (GHGs) – the strategy of 'mitigation' as it is known in the jargon of the Intergovernmental Panel on Climate Change.

To be effective against the risk of 'dangerous' climate change, mitigation will require a halving of global GHG emissions by mid century (Metz, Davidson et al., 2007). Under principles of equitable burden-sharing leading to convergence on a global *per capita* average in GHG emissions, this will imply, for the richer countries, a reduction of closer to 90 per cent by mid century, and a halving by 2030 (UK Climate Change Committee, 2008). Such a reduction strategy is equivalent to a 4 to 5 per cent annual decline in emissions every year to 2050, a change that will require major shifts in all sectors of the economy – in the energy sector, housing, transport, and food and agriculture among others. It is achievable only through determined implementation of a myriad of initiatives and policy actions: regulation, taxation, education, technology development and so forth.

The scale of the task is enormous and has caused some to question whether this level of change is realistic or worth the pain of so profound a social and economic adjustment (Gerondeau 2010) – particularly when the outcome is uncertain and the benefits are largely deferred to a future generation, which may have better technology at its disposal. However, the rationale for change in our use of fossil fuels is not only about climate change. Nearer term imperatives of addressing the impact of 'peak oil' (Wilkinson, 2008) and improved energy security, for example, are themselves strong motivations for a move away from dependence on fossil fuels. A crucial additional dimension is *current* population health.

It is becoming clear that many measures to reduce GHG emissions are likely to have a range of more immediate ancillary health benefits, which are additional to any arising in the distant future from the mitigation of climate change – the notion of the health 'co-benefits' of mitigation (Haines, McMichael et al., 2009). There are many examples. Generation of electricity from renewable sources rather than fossil fuels would help to improve outdoor air quality. More sustainable transport policies, which promote walking and cycling, could have multiple benefits arising from improved physical activity with its impact on chronic diseases, obesity, mental well-being, though potentially at some risk of road injury unless there is proper segregation of road users. Reduced consumption of animal products in high-consumption settings has the potential to impact on cardiovascular and other diseases, while also helping to reduce GHG emissions. And measures to upgrade the energy efficiency of housing have the potential to improve protection against winter cold, fuel poverty and the penetration into the home of outdoor air pollution. In lower income settings, increased uptake of cleaner-burning cookstoves could have major impact on the large global burden of illness and mortality arising from indoor air pollution relating to the inefficient and poorly ventilated burning of biomass for cooking and heating. These health benefits are direct, local and often substantial.

A new dimension to environment health and debates is therefore the recognition that many of the societal and technological changes required to meet current global environmental challenges – with their adverse ecological, social, economic and health consequences – can in large part be motivated by arguments of more immediate improvements to population health. Such gains would help offset the costs of mitigation strategies, and contribute to addressing existing global health priorities.

Thus today's environment and health priorities are framed not only by the need to respond to local hazards, but also by 'distant' global threats and the more direct opportunity to improve population health now through changes to some of the fundamentals of modern living. The type and scale of choices that face policy makers differ quantitatively if not qualitatively from those presented by the environmental concerns of previous generations. It is not clear that history has a direct analogue. Moreover, hitherto, health has not been the primary driver for change in relation to global environmental threats – but almost all can agree that protection of health is vital.

Meeting the needs of a growing and increasingly industrialised, urbanized world population, without endangering the planet's vital life-support systems and human health, presents formidable challenges for the twenty-first century. In many respects the types of challenge, which require unique levels of cooperation between states, are without precedent. But whatever paths are taken, the lessons of history should surely be an indispensible guide.

References

C. Gerondeau (2010). *Climate: The Great Delusion. A Study of the Climatic, Economic and Political Realities* (London, UK: Stacey International).

A. Haines, A. McMichael, et al. (2009). 'Public health benefits of strategies to reduce greenhouse-gas emissions: overview and implications for policy makers', *Lancet*, 374 (9707), pp. 2035–8.

A. McMichael, R. Woodruff, et al. (2006). 'Climate change and human health: present and future risks', *Lancet* 367, pp. 859–69.

B. Metz, O. R. Davidson, et al. (2007). Contribution of Working Group III to the Fourth Assessment Report of the Intergovernmental Panel on Climate Change (Cambridge, UK and New York, NY, USA, Cambridge University Press).

UK Climate Change Committee (2008). *Building a Low-Carbon Economy – the UK's Contribution to Tackling Climate Change* (London, UKCCC).

P. Wilkinson (2008). 'Peak oil: threat, opportunity or phantom?' *Public Health*, 122 (7), pp. 664–6; discussion pp. 669–70.

Index

Illustrations are indicated by page numbers in italics, Tables by bold type and Figures by "Fig."